Correlating structure and function in small molecule organic solar cells by means of scanning probe and electron microscopy

Von der Fakultät für Elektrotechnik, Informationstechnik, Physik
der Technischen Universität Carolo-Wilhelmina zu Braunschweig

zur Erlangung des Grades eines Doktors
der Ingenieurwissenschaften (Dr.-Ing.)
genehmigte

Dissertation

von
Dipl.-Phys. Michael Scherer
aus Gresaubach

eingereicht am: 12.04.2016
mündliche Prüfung am: 28.06.2016

1. Referent: Prof. Dr.-Ing. habil. Wolfgang Kowalsky
2. Referent: Prof. Dr. rer. nat. habil. Achim Enders

2016

Dissertation an der Technischen Universität Braunschweig,
Fakultät für Elektrotechnik, Informationstechnik, Physik

Correlating structure and function in small molecule organic solar cells by means of scanning probe and electron microscopy.

In this work nanoscale properties in active layers of small molecule organic solar cells are studied regarding their impact on device performance. For this, the effect of variations in stack design and process conditions is examined both electrically and with high resolution imaging techniques. Two topics are addressed: (i) the visualization of charge extraction/injection properties of solar cell contacts and (ii) the tailoring of structural properties of co-evaporated material blends for bulk heterojunction (BHJ) organic solar cells.

(i) We study the impact of controlled contact manipulation on the internal electric potential distribution of fluorinated zincphtalocyanine (F_4ZnPc)/fullerene (C_{60}) organic solar cells under operating conditions. In a detailed analytical study using photoelectron spectroscopy and in-operando scanning Kelvin probe microscopy it is demonstrated that the electric field distribution of organic solar cells at the maximum power point depends in an overproportional manner on contact properties and ranges from bulk to contact dominated even for solar cells with decent device performance.

(ii) The morphology of co-evaporated active layer blends depends on both substrate and substrate temperature. Here we study the morphology of $F_4ZnPc:C_{60}$ blends with analytical transmission electron microscopy. For all substrates used is found that co-evaporation of the materials at elevated substrate temperature (100 °C) induces a distinct phase segregation of F_4ZnPc and C_{60}. However, only when using a C_{60} underlayer, as in inverted devices, also the crystallinity of the segregated C_{60} phase increases. There is only a slight increase in crystallinity when F_4ZnPc acts as an underlayer, as typically for non-inverted devices. Solar cell characterization reveals that the crystalline C_{60} domains are the main driving force for enhanced free charge carrier generation and higher power conversion efficiencies. With this we could provide a novel explanation why record efficiencies of small molecule organic solar cells are realized in inverted device architecture only.

Zusammenhang zwischen Struktur und Funktion organischer Solarzellen basierend auf kleinen Molekülen analysiert mit Rastersonden- und Elektronenmikroskopie.

Im Rahmen dieser Arbeit werden nanoskopische Eigenschaften von aktiven Schichten organischer Solarzellen hinsichtlich ihres Einflusses auf die Solarzellen-Funktion untersucht. Auswirkungen von Variationen bezüglich Bauteildesign und Prozessbedingungen werden sowohl elektrisch als auch mit hochauflösenden Mikroskopieverfahren analysiert. Zwei Themen werden dabei adressiert: (i) Die Abbildung von Ladungsextraktions- und injektionseigenschaften an Solarzellenkontakten sowie (ii) die Anpassung struktureller Eigenschaften in koverdampften Heteroschichten aus fluoriniertem Zinkphtalocyanin (F_4ZnPc) und Fulleren (C_{60}).

(i) Wir untersuchen den Einfluss einer kontrollierten Kontakt-Variation auf die Potentialverteilung organischer F_4ZnPc/C_{60}-Solarzellen im Betrieb. Mittels Photoelektronenspektroskopie und Rasterkelvinmikroskopie wird gezeigt, dass die elektrische Feldverteilung organischer Solarzellen betrieben am Arbeitspunkt in überproportionaler Abhängigkeit zu ihren Kontakteigenschaften steht. Die Feldverteilung kann dabei selbst für verschiedene Solarzellen mit respektablen Effizienzen zwischen einer reinen Kontaktlimitierung und einer Limitierung durch die aktive Schicht variieren.

(ii) Die Morphologie koverdampfter aktiver Heteroschichten, bestimmt sowohl durch die Wahl des Substrats als auch der Substrat-Temperatur, wird mittels analytischer Transmissions-Elektronenmikroskopie untersucht. Die Koverdampfung der Materialien bei einer Substrat-Temperatur von 100 °C führt bei allen verwendeten Substraten zu einer ausgeprägten Phasenseparation von F_4ZnPc und C_{60}. Dies geht jedoch nur im Falle eines C_{60}-Substrates, üblich in invertierter Bauteil-Architektur, mit einer erhöhten Kristallinität der C_{60}-Domänen in der aktiven Schicht einher. Wird wie bei nicht-invertierten Bauteilen ein F_4ZnPc-Substrat verwendet, ist kein analoger Anstieg in der Kristallinität zu verzeichnen. Mittels Charakterisierung von entsprechenden Solarzellen zeigen wir, dass die Kristallinität der C_{60}-Domänen die Haupttriebkraft für eine effizientere Erzeugung freier Ladungsträger und einer verbesserten Solarzellen-Effizienz darstellt. Wir können damit grundlegend neu erklären, warum Rekord-Effizienzen in organischen Solarzellen basierend auf kleinen Molekülen lediglich in invertierter Bauteil-Architektur erzielt werden.

Contents

1 Introduction   1

2 Fundamentals   5
- 2.1 Charge transport in organic semiconductors 5
  - 2.1.1 Hopping transport 6
  - 2.1.2 Polarization effects on the band gap 9
- 2.2 Organic solar cells 12
  - 2.2.1 Solar cell characteristics 13
  - 2.2.2 Charge carrier separation in OPV 15
  - 2.2.3 Impact of material properties on OPV device design . 19
  - 2.2.4 The role of the built-in potential V_{bi} 20

3 Experimental details   23
- 3.1 Analytical methods 23
  - 3.1.1 Atomic force microscopy 23
  - 3.1.2 Scanning Kelvin probe microscopy 27
    - 3.1.2.1 In-operando SKPM studies 32
  - 3.1.3 Scanning electron microscopy 35
  - 3.1.4 Focused ion beam microscopy 36
  - 3.1.5 BRR integrated SEM-AFM 37
  - 3.1.6 Transmission electron microscopy 38
  - 3.1.7 Photoelectron spectroscopy 41
- 3.2 Vacuum preparation of small molecule solar cells 43
  - 3.2.1 Materials and stacks 45
  - 3.2.2 Experimentals- deposition parameter 48
  - 3.2.3 Vacuum thin film deposition 49
  - 3.2.4 Solar cell characterization 51

Contents

4 Electric potential distribution of F_4ZnPc/C_{60} small molecule organic solar cells **53**
  4.1 Electric potential distribution of the OPV stack under short circuit conditions 54
  4.2 UPS/XPS study on the hole extracting contact 61
    4.2.1 Electronic properties 64
    4.2.2 Chemical properties 66
  4.3 The influence of FIB preparation on SKPM results 70
    4.3.1 State of the art 71
    4.3.2 Experimental details 73
    4.3.3 Results of SRIM simulations: ion implantation profiles 76
    4.3.4 Results of SKPM studies: electric potential profiles 82
  4.4 In-operando SKPM studies on OPV cells with varied hole extracting contacts 85
    4.4.1 State of the art 86
    4.4.2 Experimental details 89
    4.4.3 Results: In-operando SKPM studies 90
      4.4.3.1 Results 1: Studies on illuminated devices 91
      4.4.3.2 Results 2: Studies on devices under applied bias voltages 97
    4.4.4 Discussion: Prospects and limits for in-operando SKPM studies 103

5 Structure-function relationship in F_4ZnPc/C_{60} solar cells **107**
  5.1 Bilayer solar cells with varied hole extracting contact 108
    5.1.1 State of the art 108
    5.1.2 Solar cell results on F_4ZnPc/C_{60} bilayer devices 109
    5.1.3 In-situ monitoring of bilayer thin film growth 112
      5.1.3.1 Growth and coverage studied with AFM 113
      5.1.3.2 F_4ZnPc growth studied with XPS 116
      5.1.3.3 Conclusion: in-situ monitoring of bilayer stack growth 118
  5.2 C_{60} crystallinity dictates device efficiency in $F_4ZnPc:C_{60}$ BHJ solar cells 118
    5.2.1 State of the art 119

5.2.2	Solar cell results on $F_4ZnPc:C_{60}$ BHJ devices	121
	5.2.2.1 Conventional BHJ devices	121
	5.2.2.2 Inverted BHJ devices	123
	5.2.2.3 Conclusion: solar cell results	125
5.2.3	Revealing the BHJ nanostructure with analytical TEM	126
	5.2.3.1 Phase separation of $F_4ZnPc:C_{60}$ BHJs	128
	5.2.3.2 Crystallinity in $F_4ZnPc:C_{60}$ BHJs	132
5.2.4	Prospect on the 3rd dimension-AFM studies on BHJ blends	135
	5.2.4.1 Conclusion: AFM studies on BHJ blends	138
5.2.5	Discussion: Structure-function relationship	139

6 Conclusion and outlook   143

Bibliography   147

7 Appendix   179

List of Abbreviations

AFM: Atomic force microscopy

SKPM: Scanning Kelvin probe microscopy

KP: Kelvin probe

FIB: Focused ion beam

TEM: Transmission electron microscopy

UPS: Ultraviolet photoelectron spectroscopy

XPS: X-ray photoelectron spectroscopy

UHV: Ultra high vacuum

OPV: Organic photovoltaics

OSC: Organic semiconductor

PCE: Power conversion efficiency

BHJ: Bulk heterojunction

F_4ZnPc: Fluorinated zincphtalocyanine

C_{60}:   Buckminster fullerene

RT: Room temperature

HOMO: Highest occupied molecular orbital

LUMO: Lowest unoccupied molecular orbital

CT: Charge transfer

1 Introduction

The establishment of inorganic semiconductor technology is the most important technical milestone set in the 20th century. Inventions based on this technology comprise solar cells, light emitting diodes (LEDs) and especially field effect transistors (FETs). Information technology based on silicon FETs gave rise to a second industrial revolution, their cheap manufacturing costs lead to the introduction of logical elements in electronic devices of practically all scales.

However, fabrication and purification processes for these materials are often complex and energy consuming. Because of this, high efforts in terms of fundamental science as well as R&D are expended to establish organic electronics as a complementary technology to inorganic semiconductors. Organic electronics promises the fabrication of electronic devices on arbitrary substrates at large areas and in environmental friendly and cheap high-throughput processes at low temperatures.

Already in the early 20th century first electrical studies on anthracene crystals were performed [1,2]. In the 1960s electroluminescence in organic crystals was discovered [3,4], leading to the realization of the first organic LED (OLED) in 1970 [5]. Studies by Heeger, MacDiarmid and Shirakawa lead to the fundamental understanding of the electric properties of conjugated carbon bonds and expanded the tool kit of organic electronics [6]. Their research was awarded with the Nobel Prize in chemistry in 2000 [7]. In the mid 1980s first device applications based on organic semiconductors were realized that reached performances in the order of their inorganic counterparts [8–10]. Since then there is growing interest in the realization of organic electronic devices that compete or even outperform the established technologies. Today, many products based on OLEDs are launched in the markets successfully. The realization of true color contrast due to active color pixels and low energy consumption make OLED the technology of choice for displays in

1 Introduction

mobile devices and high definition television screens. This is not (yet) the case for organic photovoltaics (OPV), although there is a steady increase in power conversion efficiency (PCE) to 13.2% for research cells and 7-8% in modules today [11] and a wide range of design options as tuning of color and transparency. The ease of applicability of integrated OPV on the one and the huge potential for tailoring the device properties by manipulation of the individual molecules on the other hand make OPV a both fascinating and promising technology. However, a deeper understanding of the fundamental processes present in organic material compounds is crucial on the road to highly efficient OPV.

Charge carrier generation and transport properties in OPV are a topic of ongoing research. The absence of long-range molecular order in the OPV systems necessitated the rethinking and -formulation of the terminology known from inorganic PV. In this process some properties of OPV devices were widely discussed but hardly accessed in experiments. Two of these properties are addressed in this work and correlated to OPV device performance, namely the electric potential distribution and the active layer morphology of OPV devices. This became possible by the application of state-of-the-art analytics with unprecedented lateral resolution to actual OPV devices.

In this thesis we apply scanning Kelvin probe microscopy (SKPM) to the cross sections of organic solar cells under working conditions to study their electric potential distribution on the nanoscale. After the first realization of such experiments at InnovationLab (iL) in 2012 and the characterization of different aspects in solution-processed OPV by Rebecca Saive [12,13], here we address the electric potential distribution in SKPM studies on vacuum processed OPV devices. The vacuum processing allows the preparation of highly defined OPV samples and the correlation of SKPM results with results from the complementary methods X-ray/ultraviolet photoelectron spectroscopy (XPS/UPS), infrared spectroscopy and scanning probe microscopy available at the integrated ultrahigh vacuum system at iL. Here we studied OPV devices with fluorinated zincphtalocyanine (F_4ZnPc)/fullerene (C_{60}) planar heterojunction active layers under working conditions with SKPM. We correlated results on solar cells with varied hole extracting contacts to device performance and findings from the above mentioned complementary methods and could demonstrate that the impact of the contact properties on the electric potential distribution of OPV devices under working conditions is significantly stronger than expected from electrical and UPS studies on non-operating devices.

The concept of the active layer morphology comprises all aspects regarding the composition of OPV active layers, namely the spatial distribution of different material components as well as their structural appearance in terms of short- and long-range ordering. Here, we access the active layer morphology with studies based on analytical transmission electron microscopy (TEM). In vacuum processed small molecule OPV, substrate heating during the application of the active layers is a popular method to control the active layer morphology due to enhancing the diffusivity of the molecules by supplying additional thermal energy. However, the specific impact of substrate heating on the active layer morphology as well as on the device performance is still a matter of debate. Some publications report on a clear improvement [14–18], others on no change [19, 20] or even on a decrease [21, 22] in device efficiency due to substrate heating. Here, we demonstrate the effect of substrate heating exemplary on OPV devices with intermixed active layer blends of the well-known material system F_4ZnPc/C_{60}. For this, we process devices as well as TEM samples mimicking devices in different solar cell architectures and on different substrate temperatures. Via the correlation of results from analytical TEM and electrical measurements we identified enhanced fullerene crystallinity induced by substrate heating as the crucial parameter for the improvement of device efficiency. This fullerene ordering depends critically on the sequence of active layer application, i.e. on whether the devices are realized in inverted or non-inverted architecture.

This work was performed in cooperation of the Technische Universität Braunschweig (sample preparation, electrical characterization and SKPM studies) and Universität Heidelberg (TEM studies) consolidated at the InnovationLab Heidelberg GmbH. The main research activities are conducted at the InnovationLab GmbH, which is a joint transfer platform allowing to merge substantial competences such as synthesis, simulation, analytics, device physics and printing of organic electronics.

1 Introduction

Outline

In the following, a brief introduction in the fundamentals of organic semiconductors with the main focus on organic solar cells is given. In chapter 3 the analytical methods and the process technologies used in this thesis are introduced. In chapter 4 a detailed study on the electric potential distribution within organic solar cell devices is presented. SKPM studies are correlated with XPS/UPS and KP measurements as well as with results from simulations. At the end of the chapter the limits and prospects of SKPM studies on operating organic electronic devices are discussed. In the following chapter 5 we present results on the structure-function relationship of both planar and intermixed bulk heterojunctions. For this, OPV device performance is related to morphology and topography of the active layers of the cells. The importance of fullerene crystallinity is discussed in the light of our findings and latest literature reports. Finally, in chapter 6 the results of this thesis are reflected and a brief outlook on potential future research topics is given.

2 Fundamentals

In this chapter we give a short review on the basic theoretical concepts used in this work. Features of charge transport in organic solids are discussed. We comment on the strong polarization effects in organic solids, which lead to a distinct dependence of semiconductor band gaps from their environment. The fundamentals of organic solar cells are discussed in the framework of general solar cell theory.

2.1 Charge transport in organic semiconductors

When describing charge transport in organic semiconductors weak intermolecular bonds and high polarizability necessitate some modifications to traditional models applied in semiconductor research. Here, we address bulk phenomena as density of states and the theory of hopping charge transport in organic solids before we discuss how polarization effects give rise to a more detailed definition of the semiconductor band gap.

The physics discussed here is based on "Organic molecular solids" from Marcus Schwörer and Hans Christoph Wolf [23].

In organic semiconductors (OSCs), opposite to inorganic semiconductors as silicon or GaAs, some of the properties usually related to semiconductors such as optical band gaps are not induced via crystal formation, but are present already in the individual molecules. Strong intramolecular bonds lead to a splitting of the valence orbitals creating an energy gap between the highest occupied molecular orbital (HOMO) and the lowest unoccupied molecular orbital (LUMO) in the individual molecular unit. There are organic solids that exhibit distinct crystallinity and long-range band transport (realized in transistors [24,25] and solar cells [18]), but the overmost systems discussed here are strongly influenced by their disordered nature. The weak van der Waals bonds imply that there is only weak overlap of the electron wavefunctions on the intermolecular scale, leading to low intrinsic charge carrier densities n_{intr} and poor electric polarizability/screening capabilities in these materials (relative permittivity $\varepsilon_r \approx 3-4$ in most organic materials).

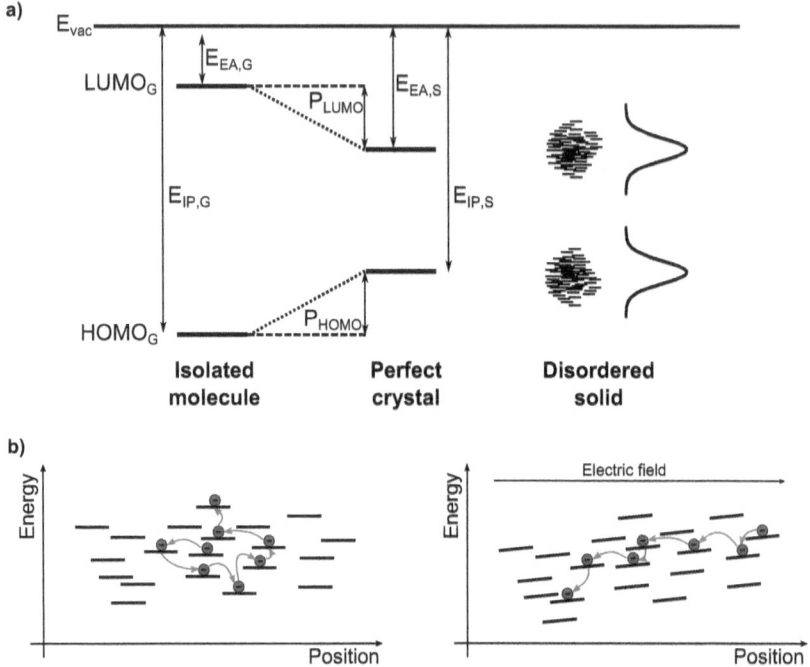

Figure 2.1: a) Energetic position of (i) isolated molecules in the gas phase (G) and condensed molecules in a (ii) crystalline and (iii) disordered solid (S). The polarization of the environment leads to an energy level shift $P_{\text{HOMO, LUMO}}$ of the condensed molecules. In the perfect crystal the polarization of the surrounding is identical for every molecule (constant $P_{\text{HOMO, LUMO}}$), in the disordered solid it varies slightly from molecule to molecule (varying $P_{\text{HOMO, LUMO}}$), leading to a Gaussian DOS. b) Hopping transport in a disordered solid. If an electric field is applied, a net current flows.

2.1.1 Hopping transport

The weak van der Waals bonding implies that no long-range order is established and that the polarization environment for all individual molecules is slightly different, leading to differing polarization modifications for all individual molecules. This gives rise to a Gaussian distributed density of states (DOS) in the organic solid, depicted in figure 2.1. The absence of long-range electron coupling implies that the charge carriers are localized on single molecules and propagate in the solid via hopping: charge carriers have to overcome certain energy barriers when moving between neighboring molecules, as depicted in figure 2.1.

Temperature dependence of the mobility

Because of the low screening in organic solids, an additional charge carrier alters the energy levels of the hosting molecule significantly. The molecule as well as its surrounding relaxes, the charge carrier becomes localized. Because of this, the mobility edges of the charge carriers are not given by the maxima of the Gaussian DOS, but are lowered (elevated) by the energy barrier Δ with respect to the LUMO (HOMO) maximum for electrons (holes) [1]. This energy barrier has to be overcome, implying that the mobility of the charge carriers in these solids is temperature-activated: $\mu \sim \exp\left(-\frac{\Delta}{k_B T}\right)$. The barrier is given by $\Delta = \frac{\sigma^2}{k_B T}$ with the variance σ^2 of the Gaussian distribution. This yields for the temperature dependence of the charge carrier mobility in disordered solids:

$$\mu \sim e^{-\left(\frac{\sigma}{k_B T}\right)^2}. \tag{2.1}$$

Field dependence of the mobility- the Poole-Frenkel effect

Another feature of disordered solids is the electric field dependence of the charge carrier mobility, described by the Poole-Frenkel effect. As discussed above the mobile charge carriers that propagate by hopping in the solid are trapped at their hosting molecule by the (average) energy barrier Δ (plus a possible additional energy barrier to the neighboring molecule). If an electric field is present in the solid, this barrier is lowered in direction of the

---

[1] We remark here that it is this localization which causes the hopping transport, *not* (only) the Gaussian DOS. Even in a weakly interacting perfect crystal this strong localization would induce hopping transport.

2 Fundamentals

field. Assuming an image charge accounting for the polarization, the field dependence of the hopping mobility is given by

$$\mu = \mu_0 \cdot \exp\left[\frac{e^{3/2}}{2k_\mathrm{B}T}\sqrt{\frac{F}{\pi\varepsilon_0\varepsilon_r}}\right], \quad (2.2)$$

with the zero-field mobility μ_0, the elementary charge e and the applied electric field F.

In the very thin organic electronic devices the exponential Poole-Frenkel factor has a strong impact on the charge carrier mobilities under device operation. For a device with an active organic layer of 100 nm the mobility increases with respect to its zero-field mobility by a factor of 6/13/38/310 if a bias voltage of 0.5 V/1 V/2 V/5 V is applied (assuming a constant electric field).

The Poole-Frenkel effect was published in 1938 [26] to describe conductivity in insulators under high electric fields and is thus not restricted to organic semiconductors. However, because of the very thin active layers (corresponding to high electric fields) present in organic electronic devices here the impact of the Poole-Frenkel effect is very prominent.

Space charge limited currents- the Child law

Organic solids are characterized by very low intrinsic charge carrier densities n_intr. Therefore in organic electronic devices under operation typically the charge carrier density in the organic active layer is governed by charge carriers injected from high quality contacts rather than by the intrinsic charge carriers, i.e. $n_\mathrm{intr} \ll n_\mathrm{inj}$. In this situation the charge neutrality in the device is canceled and the injected charge carriers create an additional, local electric field that has to be considered in the drift-diffusion equations. Assuming field-independent mobility μ, the Child law is obtained:

$$j = \frac{9}{8}\epsilon_r\epsilon_0\mu\frac{V^2}{d^3}, \quad (2.3)$$

with the applied voltage V and the thickness of the active layer d.

The Child law was proposed in 1911 [27] to describe charge transport in vacuum tubes. Because the applied bias voltage is screened by the (injected) space charge, this scenario is described as space charge limited current (SCLC).

Monte Carlo simulations- the Bässler model

The growing interest in organic electronics lead to an increasing demand for exact models describing hopping transport in organic materials. In 1993, Heinz Bässler simulated the hopping process in solids with Monte Carlo methods in the framework of the Miller-Abraham model: upward hopping is thermally activated ($\sim \exp(-\Delta/k_B T)$), downward hopping has probability 1. For the charge carrier hopping mobility he obtained:

$$\mu = \mu_0 \cdot \exp\left[-\left(\frac{2\sigma}{3k_B T}\right)^2 + C\left(\left(\frac{\sigma}{k_B T}\right)^2 - \Sigma\right) \cdot \sqrt{F}\right], \qquad (2.4)$$

with the zero-field mobility μ_0, the variance σ of the Gaussian distribution, the spatial disorder Σ, a scaling factor C and the applied electric field F [28].

The Bässler model contains both the temperature and the Poole-Frenkel field dependence discussed above.

2.1.2 Polarization effects on the band gap

In the previous section we discussed how polarization effects lead to a localization of charge carriers on hosting molecules and to relaxation of the molecular surrounding. In this section we comment on the impact of these processes on energy levels and band gaps. We give a short overview on the relaxation processes involved and discuss their influence on OSC band gaps determined with different characterization techniques such as photoelectron and UV-Vis spectroscopy.

This section is based on considerations discussed by Eric Mankel in his PhD thesis [29].

Three main contributions are involved in the molecular relaxation process:

Electronic relaxation: The valence electron configuration of the hosting molecule reacts on the presence of an additional charge carrier and relaxes

2 Fundamentals

into a new equilibrium. These fast electronic processes occur on the order of 10^{-16} s and have a high energetic contribution on the order of 1 eV.

Molecular relaxation: The new electronic environment affects also the molecular structure of the hosting molecule, and intramolecular bond lengths and angles are altered slightly. The molecular relaxation occurs on the order of 10^{-14} s and has an energetic contribution on the order of 200 meV.

Lattice relaxation: Neighboring molecules react on the discussed processes induced by an additional charge carrier and the lattice is distorted. The lattice relaxation occurs on timescales $< 10^{-14}$ s and has an energetic contribution to polarization on the order of 10 meV.

Band gaps in organic materials

On the basis of the different energies and timescales involved we discuss the energy band gaps determined in different measurements. In figure 2.2 the energy positions of the band gaps discussed in the following are given.

The adiabatic band gap is the equilibrium energy band gap of the molecule in absence of any additional charge carrier. It is not accessible in experiment.

The photoemission band gap is the band gap determined with inverse photoemission spectroscopy (IPES) and ultraviolet photoemission spectroscopy (UPS). As discussed in section 3.1.7, in the (inverse) photoemission process an electron (photon) is emitted from the solid and its energy is detected. The photoemission process occurs on a timescale of about 10^{-14} s, so it certainly includes the (large) contribution of electronic relaxation and excludes the lattice contribution. Thus the photoemission band gap is smaller than the adiabatic band gap. Because they evolve on similar timescales, it is hard to tell whether the molecular relaxation is concluded in the photoemission band gap or not. Probably there is no general answer here, but this can vary from solid to solid depending on the molecular size and details of the intramolecular bonds.

2.1 Charge transport in organic semiconductors

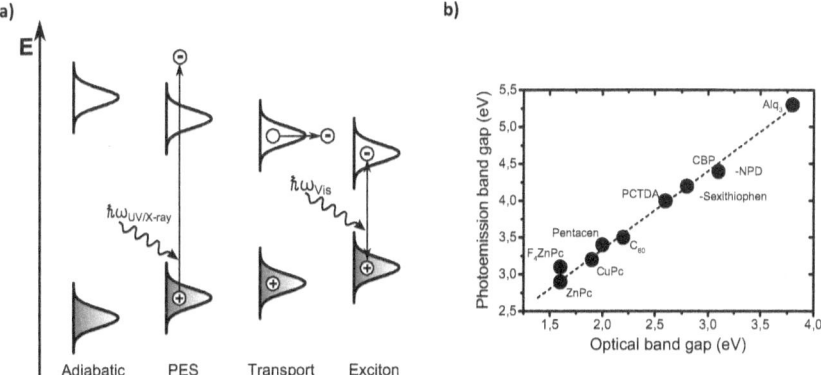

Figure 2.2: a) The different band gaps of organic semiconductors. The adiabatic band gap corresponds to the band gap of the non-ionized molecule in thermal equilibrium. The band gap decreases, depending on type and lifetime of the associated excitation. b) Comparison between bands gaps determined optically (x-axis) and via the maxima of the UPS/IPES resonances (y-axis) for materials often used in organic electronics. The band gaps of CBP and α−NPD overlap. There is a linear correlation between PES and optical band gaps and an energy offset of $E_0 = 0.97$ eV (y-intersection). For the origin of data see the end of this chapter.

The transport band gap is given by the energetic distance between the transport levels found for single electrons and single holes propagating in the solid (molecules charged one fold). The charge transport is a slow process, so all the contributions discussed above are certainly involved here. However, because of the small energy contribution of lattice relaxation, the transport band gap can be approximated by the photoemission (PES) band gap.

The optical (excitonic) band gap is given by the electronically neutral excitonic state where electron and hole are coulomb-bound on the same molecule. The optical band gap can be determined from Tauc-plots based on data from UV-Vis measurements. The exciton lifetime (typically on the ns order) is significantly higher than the timescales of molecular relaxations, so that the excitonic state contains all contributions discussed. Although its origin is still under discussion, a linear correlation between the optical and the PES band gap is found for many materials used in organic electronics.

2 Fundamentals

This is demonstrated in figure 2.2, where the PES band gaps determined from UPS/IPES spectra are plotted versus the optical band gaps determined from absorption measurements for materials widely used in organic electronics.

Implications on the notation

Because the PES band gaps vary only slightly from the transport band gaps, in this work we use band gaps determined in PES measurements when energy band diagrams or transport properties of the materials are discussed. The HOMO values given in this work were derived from the *onset* of the HOMO resonance in the UPS spectra. Optical properties are discussed in the light of optical band gaps.

Data on the PES and optical band gaps were obtained for:

- ZnPc, CuPc, PTCDA, Alq$_3$, α-Sexithiophen and Pentacen from Eric Mankel [29].

- α-NPD from [30] (PES) and [31] (optical). We added here 400 meV to the PES band gaps given by Kröger et al., because the onset of the Gaussian DOS was used for the determination of the band gaps, not the maxima as done by Mankel.

- CBP from [30] (PES) and [32] (optical). Also here we added 400 meV for the same reason.

- C_{60} and F_4ZnPc from [33] (PES) and own optical characterizations.

2.2 Organic solar cells

Solar cells convert the electromagnetic energy of photons to electrical energy. In both organic and inorganic solar cells photons with sufficient energy are absorbed in the active material of the cell and electrons are excited to the LUMO (conduction band) with a hole remaining in the HOMO (valence band). However, the mechanisms of electron-hole separation and charge extraction vary significantly in classic inorganic PV and OPV. We will discuss these mechanisms in the light of their limitations on OPV efficiency. When

2.2 Organic solar cells

comparing record organic with (more efficient) Perovskite solar cells, we find that the main obstacle on the road to higher efficiencies is the lower voltage extractable in OPV. Therefore the issue of voltage losses and possible improvement paths is discussed here in some detail.

We first introduce the common terminology used in the field of PV. On this basis the characteristics of organic PV regarding charge carrier separation and material properties are discussed and device design criteria derived. In the end some reflections on the role of the built-in potential in OPV are discussed.

Section 2.2.1 is mainly based on " Physics of Solar Cells" from Peter Würfel [34], section 2.2.2 is based in large parts on the review " Interfacial Charge Transfer States in Condensed Phase Systems" from Koen Vandewal [35]. The physics discussed in section 2.2.3 can be found for example in the review "Organic photovoltaics" from Kippelen and Brédas [36].

2.2.1 Solar cell characteristics

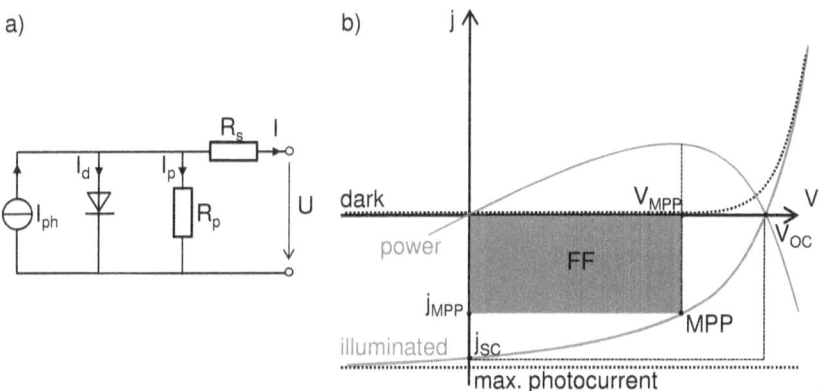

Figure 2.3: Solar cell characteristics. a) Equivalent circuit in the picture of the single diode model. Current source and diode represent an ideal solar cell, parallel (R_p) and series resistor (R_s) account for imperfect diode behavior and finite resistance of active layers and contacts. b) IV curve of a solar cell in the dark (black, dotted line) and illuminated (red). The solar cell parameter j_{sc}, V_{oc}, fill factor (FF) and maximum power point (MPP) are indicated. The maximum power output is marked in blue. Figure reprinted from [37].

2 Fundamentals

The solar cell performance is probed with current-voltage (IV) characteristics. Figure 2.3 a) depicts an equivalent circuit of an idealized solar cell neglecting the voltage dependence of parallel and series resistances (given in inorganic PV in very good approximation). The circuit is driven by the current source converting solar energy into an electrical current, the direction of the current is opposite to the polarity of the circuit. The finite parallel resistance accounts for losses induced by the imperfection of the diode, for example leakage currents through the device or a voltage-dependent charge carrier extraction. The series resistance accounts for the finite conductivity of the active materials and the sheet resistance of top and bottom contact of the cell. In an ideal cell, R_p is infinitely high and $R_\mathrm{s} = 0$.

Figure 2.3 b) depicts a non-ideal IV curve of a solar cell. The non-ideality is manifested in the finite parallel resistance R_p which leads to a non-vanishing slope at the y-axis intersect, and in the non-vanishing series resistance R_s, leading to a finite slope at the x-axis intersect:

$$R_\mathrm{p}^{-1} > 0 \implies \frac{dj}{dV}(V=0) > 0,$$

$$R_\mathrm{s} > 0 \implies \left[\frac{dj}{dV}(V=V_\mathrm{oc})\right]^{-1} > 0.$$

The terminology of solar cell characteristics is depicted in figure 2.3 b):

The short circuit current j_sc of a solar cell is given by the current flowing through the cell at $V = 0$, i.e. with short-circuited contacts. In good approximation, the short circuit current of a cell depends linearly on the illumination intensity.

The open circuit voltage V_oc of the solar cell is the voltage present between the poles if no current is flowing, i.e. if the circuit is open. In classic inorganic pn-junction solar cells the open circuit voltage is correlated with the built-in potential induced by the band bending at the junction. In OPV, its origin is more complex and discussed in detail in the next section.

2.2 Organic solar cells

The fill factor FF is given by the ratio of the maximum power extractable from the cell (blue square) and a rectangle with the borders j_{sc} and V_{oc}:

$$FF = \frac{P_{MPP}}{j_{sc} \cdot V_{oc}}.$$

The FF is limited by finite R_p and non-vanishing R_s.

The power conversion efficiency η or PCE is given by the ratio of extracted electrical energy and incoming solar energy. In the framework of the discussed parameters it amounts to:

$$\eta = FF \cdot \frac{V_{oc} \cdot j_{sc}}{P_{in}}.$$

2.2.2 Charge carrier separation in OPV

Figure 2.4 a) illustrates the mechanism of free charge carrier generation in organic solar cells. As discussed in the previous section 2.1, organic materials are characterized by low relative permittivities of $\varepsilon_r \approx 3-4$, leading to low screening in these materials. Besides that, the effective masses of electrons and holes are very high. Therefore the excitons generated in the photoabsorption process are strongly bound (0.1 eV-1 eV) and located on one single molecule (Frenkel excitons). Because of these high binding energies, the Frenkel excitons can not be separated thermally, but additional energy has to be provided. This additional energy can be generated at interfaces between (electron) donor (D) and acceptor (A) molecules with appropriate energy levels. Here, in an efficient and very fast process on the timescale of 10-100 fs the electron located in the LUMO of the donor is transfered to the (lower lying) LUMO of the acceptor, and the energy necessary to break the exciton bond is provided chemically by the difference in the LUMO energy levels of donor and acceptor (step 1 in figure 2.4 a)). However, although electron and hole are now located at different molecules, they are still bound via Coulomb interactions, the so-called charge transfer (CT) state is formed (picture 2 in figure 2.4 a)).

Only very recently it became clear that relaxed CT states indeed represent the most prominent generation channel for free charge carriers in efficient

2 Fundamentals

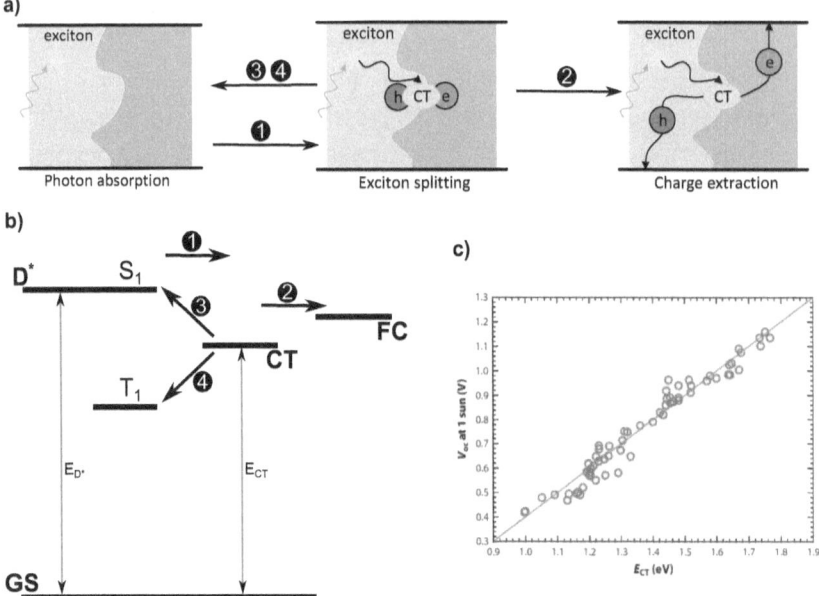

Figure 2.4: a) Schematic representation and b) energy diagram of the process of photon absorption and charge carrier separation at the donor-acceptor (DA) interface. The photon is (typically) absorbed in the donor, the generated exciton propagates to the DA interface, where it is separated via a charge transfer (CT) state. The energy diagram in b) illustrates the energy levels of the ground state (GS), the exciton hosting donor state D^*, the CT state and the state of the free charge carriers (FC). The CT state is coupled to the ground state. c) Data points of V_{oc} and E_{CT} for many OPV devices. There is a linear dependence of V_{oc} on the CT energy E_{CT} and an energy offset of $\approx 0.6\,\text{eV}$. Figure b) (c) adapted (reprinted) from [35].

OPV [35, 38–42]. However, this means that the CT states provide also a strong loss channel in the device: on the one hand, the limited lifetime of the CT state ($\sim$ ns) and the existing loss channels (step 3, 4 in figure 2.4 a)) imply that the already separated electron-hole pairs can still recombine in a geminate manner [2]. On the other hand, CT states provide also a non-geminate recombination channel for free charge carriers (exploited in OLEDs).

We want to remark here that the popular picture of the effective OPV band gap $E_{\text{gap}}^{\text{eff}}$ ($= E_{\text{HOMO}}^{\text{D}} - E_{\text{LUMO}}^{\text{A}}$) that dictates the maximum V_{oc} is still valid, because the energy level E_{CT} of the CT state scales with $E_{\text{gap}}^{\text{eff}}$ for the overmost part of the DA material combinations used in OPV. The evidence for the impact of $E_{\text{gap}}^{\text{eff}}$ on V_{oc} is striking [43] and beyond controversy in the OPV community. Here we zoom into this picture to get a grasp on possible trends of next generation OPV.

The role of the CT state in free charge carrier generation

In a detailed balance approach the properties of the charge transfer state as a free charge carrier generation and loss channel are connected [44]. Balancing all generation and recombination processes and assuming free charge carrier generation via relaxed CT states only yields for the open circuit voltage V_{oc}:

$$qV_{\text{oc}} = E_{\text{CT}} + k_{\text{B}}T \cdot \ln\left(\frac{G}{k_{\text{B}}N_{\text{CT}}}\right) \approx E_{\text{CT}} - 0.6\,\text{eV} \qquad (2.5)$$

with the CT energy E_{CT}, the free charge carrier generation rate G and the density of CT complexes N_{CT}. The ln term is < 0 here, the difference of $0.6\,\text{eV}$ between V_{oc} and E_{CT} is found empirically.

In figure 2.4 c) the dependence of V_{oc} from E_{CT} is plotted for state-of-the-art OPV, containing both polymer-fullerene and small molecule solar cell devices. V_{oc} is found to differ from E_{CT} by about $0.6\,\text{eV}$. Of course, the transition from D^* to the CT state is driven by a certain energy gradient. However, it was found that an offset of about $100\,\text{meV}$ is sufficient here. This means that for the current generation OPV the largest part of the collective loss in V_{oc} of $0.6\,\text{eV}$ is attributed to the difference in V_{oc} and E_{CT}. Rewriting

---

[2] Recombination is called "geminate" if the involved electron-hole pair was generated in the same absorption process.

2 Fundamentals

equation 2.5 points towards some possible strategies to decrease this loss:

$$E_{\text{CT}} - V_{\text{oc}} = k_{\text{B}}T \cdot \ln\left(\frac{k_{\text{rad}}N_{\text{CT}}}{G}\right) + k_{\text{B}}T \cdot \ln\left(\eta_{\text{EL}}^{-1}\right) \qquad (2.6)$$

with the radiative part k_{rad} of the CT decay rate, the free charge carrier generation rate G, the CT complex density N_{CT} and the internal electroluminescence (EL) efficiency η_{EL}. Both ln terms are < 0. This equation can be summarized in the sentence: "An ideal solar cell has to be an ideal OLED too".

EL efficiency is given by the ratio of the radiative and the non-radiative CT decay rate: $\eta_{\text{EL}} = k_{\text{rad}}/k$ and typical values for DA blends containing fullerenes are in the range of 10^{-6}, so that the second term accounts already for about 360 mV of loss in V_{oc}.

Some strategies to fundamental progress in OPV device efficiency could thus be:

- Tailoring the density of charge transfer complexes N_{CT}. Concerning N_{CT} there is a trade-off between providing sufficient DA interface area for exciton splitting and introduction of recombination centers.

- Reduction of the radiative decay rate k_{rad}. This is probably the hardest task, because knowledge on the non-radiative pathways is rare. Besides that, a reduction in k_{rad} would affect also η_{EL} ($\eta_{\text{EL}} \sim k_{\text{rad}}$).

- Reduction of nonradiative decay paths to increase the electroluminescence efficiency η_{EL}. This is probably the most promising road towards OPV with high V_{oc}. It was demonstrated that cells with high $V_{\text{oc}} \approx 1.33\,\text{eV}$ exhibiting losses of only $E_{\text{CT}} - V_{\text{oc}} \approx 0.37\,\text{eV}$ can be realized using non-fullerene acceptor materials [45].

- A complementary approach to high-V_{oc} devices is the realization of organic solar cells with a cascade-energy-level-alignment [46, 47]. In these devices a two-step exciton dissociation process is employed which does not comprise the CT state. With this, even higher V_{oc} exceeding E_{CT} would be possible in solar cells with highly ordered active layers.

2.2.3 Impact of material properties on OPV device design

In this section we give an overview on the properties typically found in OPV materials and how they dictate the architecture of state-of-the-art OPV devices. On the end of every paragraph the derived design rules are highlighted.

Photon absorption and free charge carrier transport

The first step in PV energy conversion is the photon absorption. The absorbing (typically donor) materials used in OPV exhibit very strong absorption coefficients in the order of 10^5 cm^{-1}, so that already with very thin active layers in the ten nm range a significant share of the sunlight can be harvested. This is not surprising considering that many of these materials were actually developed as dyes (or are derivatives of dye materials). The efficient absorption in thin layers is crucial on the other hand because the charge carrier mobilities in organic materials are very low. For example, the donor material F$_4$ZnPc used in this work has a hole mobility $\mu_{F_4ZnPc} \approx 10^{-5}$ cm^2(Vs)$^{-1}$ which is about 8 orders of magnitude lower than that of crystalline silicon ($\mu_e^{Si} \leq 1400$ cm^2(Vs)$^{-1}$, $\mu_h^{Si} \leq 450$ cm^2(Vs)$^{-1}$). Because of this, there is a trade-off concerning the active layer thickness between sufficient absorption on the one and charge extraction from the active layer on the other hand. Therefore OPV can be realized in a thin film architecture only.

- Very thin active layers (~ 100 nm) allow sufficient absorption and short free charge carrier transport paths to the electrodes.

Charge carrier separation

The charge carrier separation necessitates the presence of appropriate electron donor and acceptor materials in the active layer of an OPV device. This was first found and exploited systematically by Tang in 1986 [8] with organic solar cells in a DA bilayer architecture, i.e. a planar layer structure of the D and the A material. Heeger et al. [48] proposed an intermixed DA structure, the bulk heterojunction (BHJ), in 1995. The BHJ is a self-organizing, intermixed DA blend that optimally provides (i) sufficient DA interface area for efficient

Frenkel exciton dissociation, (ii) no excess DA interface providing CT states acting as recombination center and (iii) interconnected percolation paths for efficient extraction of free charge carriers. The exciton diffusion length in organic materials ranges from 3 to 40 nm, so typically a phase separation of the D and the A materials in this range is aimed. Many reports and also studies presented in this work emphasize the importance of crystallinity of the phase segregated D and A domains, because it allows for a larger delocalization (larger separation) of the charge carriers bound in the CT state. The crystallization is also desirable because it leads to enhanced charge carrier mobilities.

- Thin donor and acceptor layers (< 50 nm) for cells in bilayer architecture.
- Phase segregation of 10 to 40 nm in the self-organized BHJ architecture.
- Crystallinity of the segregated phases for more efficient exciton splitting and higher mobilities.

Charge carrier extraction

Efficient OPV cells have distinct selective contacts that typically fulfill two tasks: (i) they inhibit excitons and electrons (holes) to enter the hole (electron) extracting contact, thus ensuring high V_{oc}. This is of high importance in OPV because of the thin active layers involved. (ii) They typically enhance the built-in electric field V_{bi} in the cell (discussed in detail in the next section).

- Selective contacts for high V_{bi} and good blocking properties of the contacts.

2.2.4 The role of the built-in potential V_{bi}

The role of the built-in potential V_{bi} in OPV devices is a topic broadly discussed in the OPV community. Basically, the question is whether (and how) the V_{oc} of an OPV device depends on its V_{bi}. Whereas in classic PV devices based on a pn-junction the V_{bi} exhibits the ultimate upper limit for V_{oc}, their interplay in OPV is way more complex. Here, a short overview on the discussion on the level of reviews is given. Afterwards, some studies and their most prominent results with relevance for this work are presented.

2.2 Organic solar cells

When it comes to the built-in potential V_{bi} in OPV, many review articles (especially from the early 2000s) refer to the metal-insulator-metal (MIM) model: considering the organic layer effectively charge carrier free, in the equilibrium case (0V bias voltage and no illumination) "the built-in electric field resulting from the difference in the metals' work function is evenly distributed throughout the device" (exemplary here Hoppe and Sariciftci [49]). If the cell is illuminated this drift field drives the separated charge carriers to their respective electrodes. However, neglecting the complex nature of electronic organization on metal (or transition metal oxide)/organic or organic/organic interfaces, this simple picture of a constant electric field can hardly account for real device physics. Kippelen and Brédas stated that the different electrode work functions lead to a built-in potential, resulting in an electric field that assists the transport of charges [36]. Probably not by chance, no statement is made concerning the *distribution* of this field. Also Qi and Wang, as well as Kirchartz, emphasized the beneficial effect of V_{bi} on V_{oc} in OPV [50, 51]. Kirchartz made a good point with the pragmatic statement that "the built-in voltage V_{bi} ensures that the applied forward bias V (e.g. at the maximum power point under illumination) can drop somewhere". In a very instructive review Greg and Hanna discussed the issue of V_{bi} in OPV. They emphasized the leading role of diffusion rather than drift via a built-in field as the dominant driving force in OPV [52].

In 1986 Tang introduced the modern OPV cell characterized by its exciton splitting mechanism at the organic/organic DA interface. Recalling the charge separating character of V_{bi} in pn-junction based solar cells, he related the term of the built-in potential to the DA interface rather than to the entire OPV cell. He stated that "it appears that the electrodes simply provide ohmic contact". In 1996 Campbell et al. applied electroabsorption spectroscopy to MIM devices based on polymers used in OPV [3]. They found that the built-in field in the devices scales linearly with the difference $\Delta\phi$ of the electrode work functions [54]. However, they did not relate these results to OPV performance. Brabec et al. demonstrated in 2001 that in bulk heterojunctional OPV there is only a minor influence of $\Delta\phi$ on the V_{oc} [55]. Studying high quality contacts, Mihailetchi et al. found that when using ohmic contacts, V_{oc} is governed by

---

[3] In electroabsorption spectroscopy, resonance shifts in molecular orbitals based on the Stark effect are exploited to estimate the internal electric field within a MIM diode structure [53].

2 Fundamentals

the effective DA band gap. Only in case of non-ohmic (i.e. non-optimized) contacts, V_{oc} is influenced by $\Delta\phi$ [56]. Wang et al. estimated the influence of V_{bi} on V_{oc} being less than 10%. At the same time they argued that the dark carrier recombination at the contacts limiting V_{oc} is determined by $\Delta\phi$, emphasizing its importance for V_{oc} independent of the concept of the built-in potential [57].

In recent years however, most studies agree on some main principles:

1. The upper limit for the V_{oc} of an OPV device is given by the effective DA band gap rather than by V_{bi} [17, 43, 58, 59].

2. In general, a higher V_{bi} goes along with increased PCE. However, the improved PCE is not attributed to an increase in V_{oc} only, but also j_{sc} and FF rise [51, 59, 60].

3. An increase in V_{bi} correlates with a higher V_{oc}, but the slope of this correlation is rather shallow (mostly estimated to be $0.05 - 0.3$) [57, 58, 60].

4. V_{bi} depends critically on the active materials as well as on the energy levels of the transport layers. Thus not only the $\Delta\phi$ of the electrodes, but the composition of the entire stack determines V_{bi} [59–61].

5. Interface phenomena such as dipole formation can reduce the *effective work function* of an electrode significantly [51, 61].

3 Experimental details

The bulk part of the work presented in this thesis was performed at the clustertool ultrahigh vacuum (UHV) system. The clustertool is an integrated UHV system and the heart of the analytic facility at iL. It consists of several vacuum chambers interconnected via transfer chambers and a glovebox for the transfer to vacuum under controlled atmosphere (see figure 3.1). The main benefit of the system is the preparation of ultraclean samples and their in-situ characterization with a combination of several tools. Thus it is possible to investigate the sample from very different perspectives, and to correlate between morphological, chemical and electronic properties. Here this is exploited in in-situ studies with atomic force microscopy (AFM)/Scanning Kelvin probe microscopy (SKPM) and X-ray/ultraviolet photoelectron spectroscopy (XPS/UPS) for correlation of interface band bending, chemical composition and morphology (see section 4.2 and 5.1.3). All devices presented in this work were processed at the preparation chambers in clustertool.

3.1 Analytical methods

In the following, the analytical tools and setups used in this work are introduced. First, the principles of AFM and SKPM are discussed. After some words about imaging with scanning electron microscopy (SEM) and sample preparation with a focused ion beam (FIB), the DME Zeiss BRR setup is presented, which brings together all of the aforementioned methods. Analytical transmission electron microscopy (TEM) will be followed by XPS/UPS.

3.1.1 Atomic force microscopy

In AFM the surface topography of a sample is scanned line-by-line with a very sharp tip. Reaching sub-nm resolution, it exceeds the Abbe diffraction limit

3 Experimental details

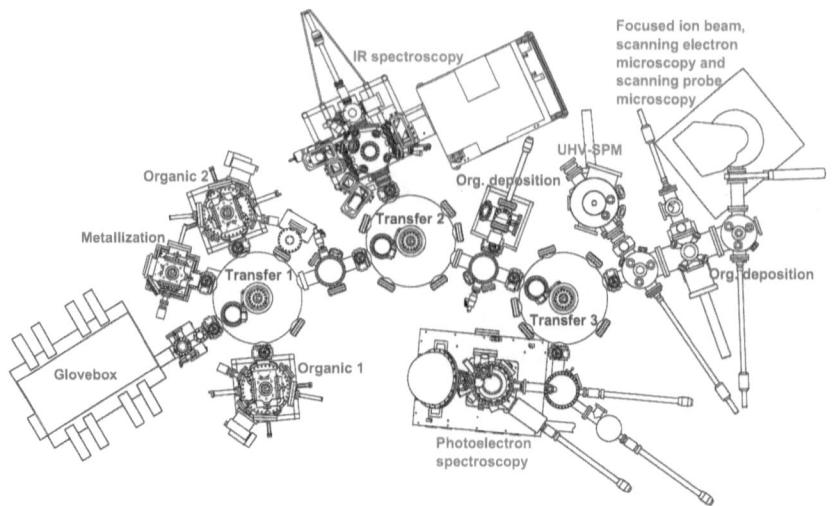

Figure 3.1: Overview on the clustertool integrated UHV system. Samples or substrates are introduced via the glovebox, organic as well as metal thin films and entire devices are processed in the preparation section connected to Transfer 1. All chambers are interconnected via the transfer chambers. This allows for the in-situ characterization of one specific sample with a manifold of surface characterization techniques.

(resolution $\geq \lambda/2$) for visible light by about three orders of magnitude. After the huge success and the fast adaptation of scanning tunneling microscopy (STM) [62] in the early 1980s, the research group at IBM Zürich around Gerd Binnig and Heinrich Rohrer could realize the first AFM in 1986 [1] [64]. Since then, this technique has seen a tremendous growth making it today the most prominent tool for imaging and manipulating matter on the nanoscale. Easy handling, applicability in various environments (from vacuum to highly adhesive liquids) and temperature regimes and the moderate acquisition costs are some of the reasons for its popularity within the scientific community [65, 66]. For a more precise treatment of the AFM physics than given here it is referred to [67] or [68].

---

[1] The same year, 1986, they were awarded the nobel prize for their invention of the STM [63].

3.1 Analytical methods

The origin of atomic forces

During the development of the STM it became evident that in addition to the tunnel current, there is further interaction between tip and sample surface, strong enough to influence the tunneling current $I_t(z)$. However, unlike the the well-understood concept of tunneling current in STM, this interaction is based on a multitude of different forces with length scales ranging from µm to sub Å. In sum these forces result in a Lennard-Jones potential. The most prominent forces in AFM are

- *van der Waals forces* caused by dipole-dipole interactions (always attractive). Assuming tip and sample consisting of an effective medium instead of individual atoms, van der Waals interactions between them can be estimated (Hamaker approach [69]). For a spherical tip with radius of curvature R in (some nm) distance z of a flat surface this approach yields

$$V_{\text{vdW}} = -\frac{A_H R}{6z}, \tag{3.1}$$

  with the (material dependent) Hamaker constant A_H ($\approx 1\,\text{eV}$ for most configurations [68]). For an AFM tip with a radius of curvature of $R = 20\,\text{nm}$ and a small tip-sample distance of $z = 5\,\text{Å}$ this yields a van der Waals energy of $\approx -7\,\text{eV}$ and a corresponding force of $\approx -5\,\text{nN}$.

- *electrostatic forces* caused by a non-vanishing electrostatic potential between tip and sample (always attractive). Applying the effective medium approach and modeling the tip as an upside down cone with a sphere on top yields:

$$F_{\text{el}} = -\pi\varepsilon_0 \frac{R}{z} \left(V_{\text{tip}} - V_{\text{sample}}\right)^2.$$

  For similar values as used above and an electrostatic potential difference of $V = 1\,\text{V}$ the electrostatic force is $\approx -1\,\text{nN}$ [70]. Obviously, F_{el} can be avoided by grounding the setup to assure $V_{\text{tip}} = V_{\text{sample}}$.

- *exchange forces* based on quantum mechanical interactions between the electron systems in tip and sample. As known from basic solid states quantum mechanics, exchange forces are attractive on the longer

3 Experimental details

and repulsive on the shorter range. Thus, they can be estimated by a Lennard-Jones potential. However, because of the very strong gradients involved meaningful estimations are hardly possible here.

Tapping mode AFM

In all the measurements presented here the AFM was operated in *tapping mode* (first introduced in 1991 by Albrecht et al. [2] [71]) with the feedback loop locked on the oscillation amplitude (amplitude modulated AFM). The cantilever is driven by a Piezo crystal with a frequency close to its resonance ω_0, leading to an excitation amplitude of typically $\lesssim 20$ nm. The approach of the cantilever to the sample surface leads to a damping of its oscillation amplitude according to Hookes law ($F \sim -x$). This damping represents the control variable for the feedback loop which keeps the tip in constant distance over the surface while scanning it line-by-line.

In frequency modulated AFM the frequency shift of the cantilever resonance caused by tip-sample interaction is used as the feedback control variable. It promises ultrahigh resolution and fast data acquisition in vacuum environments [67].

In the AFM measurements presented in this work *tapping mode* is used because

- it is less invasive than contact modes and therefore the preferred technique in the field of soft matter materials [72].

- it does not suffer from $1/f$ noise present in force detectors on the low ($f < 50$ Hz) frequency range [67]. A bandpass filter centered at ω_0 is used for signal processing, suppressing low frequency noise.

- many of our AFM measurements are accompanied by SKPM measurements. Because an external bias voltage between tip and sample is applied, these are hardly performed in contact mode. Tip-sample contact caused by snap-in effects [73] would be probable here, causing shortcuts and severe damage to tip and sample.

---

[2] Back then, they called their method „frequency modulated AFM", a term referring today to a different AFM mode locking on the phase shift of the cantilever's resonance. Today we refer to their method as amplitude modulated AFM.

3.1 Analytical methods

Technical details

In this work data from three different AFM setups are presented.

The *DualScope 95* from DME is a conventional AFM setup working in ambient air [74]. For vibration reduction, it is mounted on a granite table. It allows for high throughput AFM analysis with scan velocities of up to 130 µm/s along the fast axis, resulting in acquisition times of only 1 minute/frame for AFM images of $2 \cdot 2\,\mu m^2$ or $4 \cdot 4\,\mu m^2$ and $512 \cdot 512$ pixels (typical scan fields for AFM images presented here). For a more detailed description of the setup and its abilities (especially concerning in-operando SKPM studies) it is referred to the master thesis of Sebastian Hietzschold [75].

The *UHV-AFM* is operated in one of the characterization chambers of the clustertool (see figure 3.1) under UHV conditions of $< 10^{-8}$ mbar. Thus, vacuum processed samples can be analyzed in-situ without breaking the UHV chain. It offers the possibility of SKPM characterization, as used in the measurements in chapter 4.1. For a detailed description of this setup see [75].

The *BRR integrated SEM-AFM*, which is discussed in chapter 3.1.5.

The AFM measurements presented here were performed with Arrow cantilever from Nanoworld [76]. They have a length of 160 µm, a width of 45 µm and a thickness of 4.6 µm. Their force constant is 42 N/m, the resonance is at about 285 kHz and they have a tip radius of curvature of < 10 nm, allowing for high resolution AFM characterization.

3.1.2 Scanning Kelvin probe microscopy

Scanning Kelvin probe microscopy (SKPM) combines the setup of an AFM with that of a classic Kelvin probe (discussed in [77, 78]) to image the contact potential difference CPD (also Volta potential) between a tip and a sample surface with nm resolution [79]. After calibrating the work function ϕ_{tip} of the tip with HOPG [80], the work function of the sample is determined by

$$\phi_{\text{sample}} = \phi_{\text{tip}} - CPD_{\text{tip-sample}}.$$

In SKPM, the (conductive) AFM tip serves simultaneously as AFM and Kelvin probe: it forms a capacitor with the sample surface, and the tip-sample CPD is recorded for every pixel. Figure 3.2 shows a SKPM setup. For

3 Experimental details

a more detailed treatment of the SKPM physics than given here it is referred to [81].

SKPM modes: FM versus AM

Several setups for SKPM measurements were developed since its introduction 1991 by Nonnenmacher et al. [79]. The concepts of the two most prominent techniques, namely amplitude modulated (AM) and frequency modulated (FM) SKPM, are briefly discussed here. Afterwards, we address the physics of AM SKPM (used in this work) in more detail.

In both AM and FM SKPM the electrostatic potential between sample and conductive tip is modulated with an AC voltage ($\sim \cos(\omega_{mod}t)$), resulting in an oscillation of the cantilever with ω_{mod} (used for AM) and sidebands at $\omega_0 \pm n \times \omega_{mod}$ (used for FM). With Lock-In techniques, these oscillations are demodulated and the CPD is extracted by nulling the oscillation amplitude A at ω_{mod} (AM) or the frequency shift $\delta\omega_0$ of the cantilever resonance (FM) with a DC voltage [3]. According to Zerweck et al. [82], the FM SKPM allows for a better energy and lateral resolution than the AM method: this stems from the fact that AM SKPM probes the (long range) electrostatic force F_{el}, whereas FM SKPM probes its gradient $\frac{\partial F_{el}}{\partial z}$, which has a shorter decay length. However, others claim that the picture of Zerweck et al. is incomplete and also in FM SKPM there is significant contribution of F_{el} [84]. Because of technical issues concerning the basic electronics of our setup, we could unfortunately not realize the FM method in the scope of this work. Thus, only data gathered by AM SKPM and the respective theory are presented in the following.

Fundamentals of AM SKPM

Prerequisite for every SKPM measurement is a proper setup of the AFM system. We use tapping mode AFM on the cantilever resonance of $\omega_0 \approx$ 300 kHz and apply an additional AC voltage ($A_0 \cos(\omega_{mod}t)$) with $\omega_{mod} \approx \omega_0 - 8$ kHz as well as a DC voltage to extract the CPD of the sample. The principle of the measurement can be understood by basic capacitor physics:

---

[3] We recommend the reading of Zerweck et al. [82] and Glatzel et al. [83] for a proper understanding of the two methods.

3.1 Analytical methods

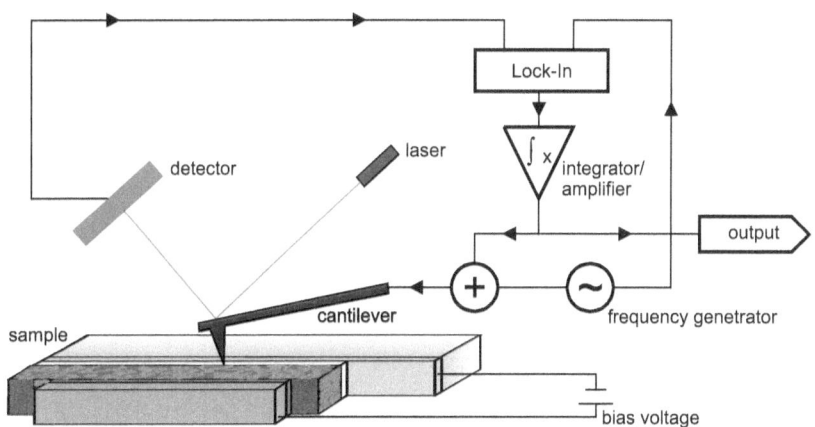

Figure 3.2: Working principle of SKPM. Parallel to an AFM measurement the CPD of the sample is imaged by using a frequency orthogonal to the one used for AFM. If a bias voltage is applied to the sample, devices can be characterized during operation. Here a cross-sectional measurement of a device with an applied bias voltage between top and bottom electrode is sketched.

If a cantilever tip is approached to a surface at differing electrostatic potential, an electric capacitor is formed. By using $dq = C du$ we obtain for the energy of this configuration

$$W = \int_{Q=0}^{Q} v dq = C \int_{v=0}^{V} u du = \frac{1}{2} C V^2, \tag{3.2}$$

so that the electrostatic force acting between cantilever and tip is given by

$$F = -\frac{1}{2} \frac{dC}{dz} V^2, \tag{3.3}$$

with the potential difference

$$V = V_{\text{applied}} - V_{\text{cpd}} = V_{\text{dc}} + V_{\text{ac}} \sin(\omega t) - V_{\text{cpd}}. \tag{3.4}$$

3 Experimental details

Insertion of equation 3.4 in equation 3.3 yields

$$F = -\frac{1}{2}\frac{dC}{dz}\left[V_{dc} + V_{ac}\sin(\omega t) - V_{cpd}\right]^2$$
$$= -\frac{1}{2}\frac{dC}{dz}\left[(V_{dc} - V_{cpd})^2 + 2(V_{dc} - V_{cpd})V_{ac}\sin(\omega t) + \frac{V_{ac}^2}{2}(1 - \cos^2(2\omega t))\right]$$
(3.5)

These forces induce an additional deflection of the cantilever tip which is detected by the laser system of the AFM setup. Rearranging equation 3.5 on the frequency domain yields $F = F_{dc} + F_\omega + F_{2\omega}$ with

$$F_\omega = -\frac{dC}{dz}(V_{dc} - V_{cpd})V_{ac}\sin(\omega t). \quad (3.6)$$

With this, for the demodulated signal of equation 3.6 holds:

$$F_\omega = 0 \implies V_{dc} = V_{cpd},$$

so that for the output V_{dc} of a feedback loop nulling the cantilever deflection at frequency ω holds $V_{dc} = V_{cpd}$. Thus, by locking the DC output of the closed feedback loop for every pixel, the CPD is imaged in a 3-dimensional manner as known from AFM.

Resolution limits in SKPM

SKPM is highly sensible towards environmental influences as well as topographical details of the sample. Therefore, the (lateral as well as energy) resolution is generally hard to predict. Before discussing the limitations of our system, a short overview on the potential of ultrahigh resolution SKPM at highly defined model systems is given.

In 2000, Kitamura et al. [85] first reported on atomic-scale resolution in a SKPM experiment under UHV conditions. They used FM methods both for AFM and SKPM signal acquisition. UHV experiments of other groups followed, and Enevoldsen et al. [86] were the first to reach atomic resolution with AM SKPM (and FM AFM). Sadewasser et al. [87] provided first theoretical and semiempirical work on the origin of atomic resolution in SKPM.

They stated that the observed features are induced by tip-surface interactions rather than being genuine properties of the freestanding surface [4]. Very recently, Perez Leon et al. [88] correlated ultrahigh resolution SKPM results with DFT calculations. They demonstrated that the tip-sample interaction is not a prerequisite for CPD contrast on the atomic scale. Their results demonstrate that SKPM is indeed capable of imaging the real energetics of surfaces with sub-nm precision.

However, all of the mentioned experiments were performed under highly defined conditions concerning environment and cleanliness of the (atomically flat) surfaces as well as the probe tips. With a setup as used in this work the lateral resolution is typically in the range of 20 nm, the energy resolution in the range of 5 mV [89].

Convolution effects in SKPM

One of the main issues constraining the resolution in AM SKPM are convolution effects between cantilever and sample. The electrostatic forces probed in SKPM are very long-range. Thus, not just the very tip but the whole cantilever contributes to the SKPM signal. Elias et al. isolated the contributions of tip and cantilever and found that the latter can attenuate the absolute CPD signal by up to 50% [90]. For the basic understanding of convolution effects in SKPM, the very instructive paper of Charrier et al. [91] is highly recommended. Simple concepts to increase the quantitative power of SKPM measurements are presented, some of them we apply to the results presented in chapter 4.4. There are several concepts available for the deconvolution of SKPM results [92, 93]. However, besides being very complex they require (i) a very good knowledge of tip and sample (which may not always be given) and (ii) assumptions concerning the symmetry of the tip-sample system (which may not be fulfilled in each measurement). Baier et al. [94] gave a user-friendly guide towards quantitative SKPM studies of nanoscale potential distributions which is worth reading. However, they refer to measurements on nanostructures only, so that the applicability to the work presented here is limited.

Besides convolution, also electrification effects, i.e. effects of surface charging

---

[4] Keeping in mind that the work function is a macroscopic concept, the question of atomic resolution in SKPM is not only of technical nature.

3 Experimental details

induced by the tapping cantilever can tamper SKPM results. Li et al. [95] demonstrated that a proper adjustment of the tapping amplitude can prevent these artifacts.

3.1.2.1 In-operando SKPM studies

Obviously, in SKPM it is crucial to care for a defined electric potential between tip and sample. In our setup, the above mentioned high frequency voltage V_{ac} is applied to the cantilever, with the sample kept grounded. As illustrated in figure 3.2 we can use an additional/independent circuit to drive a sample consisting of an electronic device and study its potential distribution in-operando, i.e. under working conditions. For vertical devices such as OPV, this is done by scanning the device cross section with the SKPM tip while applying illumination and an external bias voltage. The first groups that reported of in-operando SKPM studies on devices were Kikukawa et al. [96] from Hitachi labs as well as Vatel [97] and Chavez-Pirson [98] from NTT labs [5].

To illustrate the principles of in-operando SKPM studies, in figure 3.3 a virtual measurement on a silicon pn-homojunction is depicted. As in the measurements presented in this work, the device cross section was exposed and is scanned with SKPM. The electric potential is determined along the charge carrier transport path and single line scans are discussed under the assumption that they represent the physics of the entire device. Here, we first discuss the left hand side of this figure, referring to the real potential distribution within the device. Afterwards we address the right hand side, where the effects of CPD measurements are taken into account. The CPD of an electronic device under operation is given by the superposition of the *local work function* (here called CPD_{0V}) and the *effect* of the applied bias voltage at the respective device position (here referred to as CPD_{rel}) [6]:

$$CPD = CPD_V = CPD_{rel} + CPD_{0V}. \quad (3.7)$$

---

[5]So interestingly many of the pioneering in this field was performed at Japanese R&D labs.

[6]Obviously, for the CPD values discussed applies: $CPD = CPD(\bar{x})$, so speaking of and calculating CPD values here always implies that we refer to a specific position $\bar{x}$ within the device.

3.1 Analytical methods

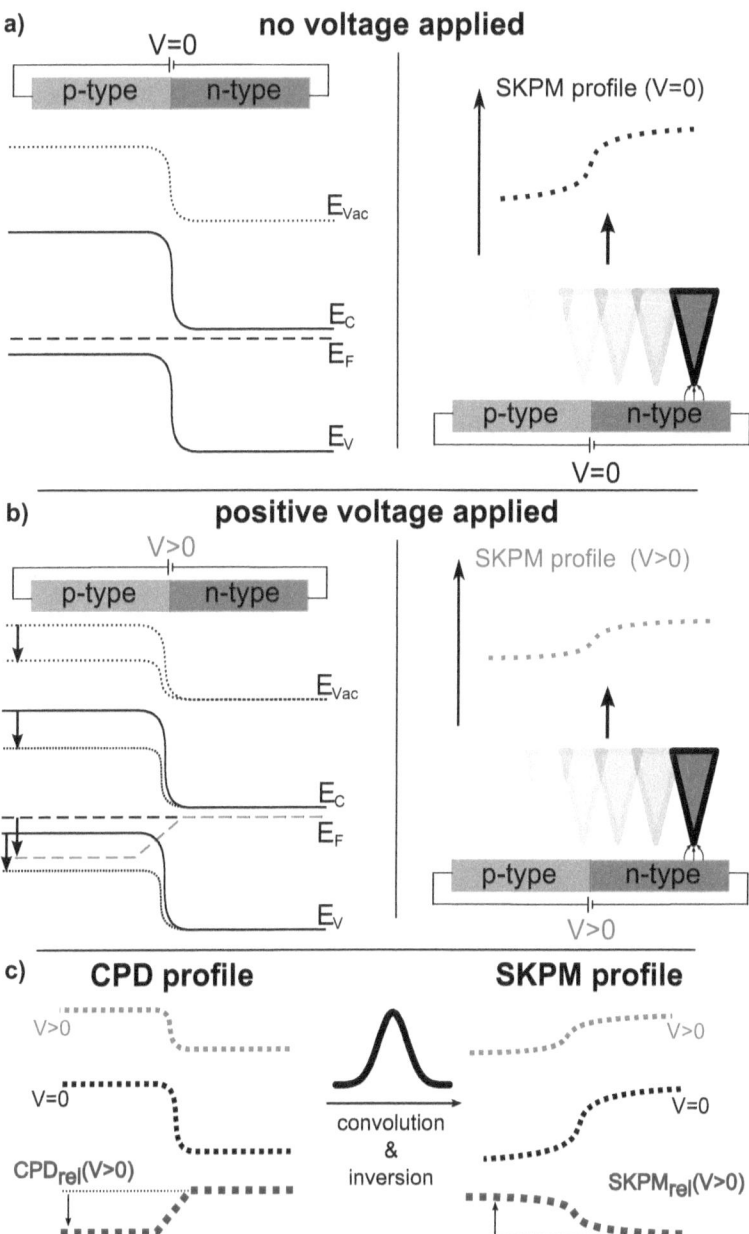

Figure 3.3: In-operando SKPM studies on a pn-junction device. a) Device under short-circuit condition. The SKPM profile represents the (convoluted and inversed) vacuum level distribution. Low values in the SKPM profile correspond to high work functions. b) Device under applied forward bias. c) In the relative profiles, the bias induced change in the profiles is highlighted.

3 Experimental details

This equation shows that the net effect of the bias voltage onto the potential distribution of the device can be obtained by subtracting CPD_{0V}. Generally, CPD_{rel} (blue profile in figure 3.3) is the parameter of highest interest when applying in-operando SKPM. In these relative profiles, obstacles in the charge transport path caused by interface barriers or material sequences with poor conductivity can be identified. In the case of the pn-homojunction discussed here, we learn from the relative profile CPD_{rel} that the bias voltage applied drops in the depletion zone right at the pn junction. However, CPD_{rel} does not tell anything about the *origin* of these obstacles. Therefore, a proper background on the device physics and architecture is an absolute must for meaningful in-operando SKPM studies. Considering actual measurements (right hand side in figure 3.3), convolution effects (discussed in the previous section) lead to a softening in the SKPM contrast when compared to CPD. Of course, this applies also to the relative profiles extracted from such measurements.

Another issue concerns the discussion of SKPM profiles in the usual electron band diagram with its negative energy axis. The SKPM profile has to be mirrored on a horizontal mirror axis to be interpreted as the vacuum level in the electron energy band diagram (compare figure 3.3). Therefore the device anode in in-operando measurements under positive bias voltage is shifted upwards instead downwards as in electron band diagrams.

Technical details

The SKPM measurements presented here are performed using conductive Pt/Ir coated ATEC-NCPt-50 cantilever from Nanosensors [99]. They have a length of 160 µm, a width of 45 µm and a thickness of 4.6 µm. Their force constant is 45 N / m, the resonance is at about 335 kHz and they have a tip radius of curvature of < 20 nm. The advantage of the ATEC architecture for our integrated AFM-SEM system is the tip visibility from above. With this, we can position the tip of the cantilever under SEM observation with very high precision ($\approx$ 1 µm) right at the area of interest (for the system see section 3.1.5).

3.1.3 Scanning electron microscopy

Soon after the proclamation of the wave character of matter by de Broglie in 1924, scientists thought about electron imaging to overcome the limitations of optical microscopy. This approach is based on the much shorter wavelength of accelerated electrons compared to optical photons. The de Broglie wavelength of electrons accelerated with a voltage V is given by $\lambda = \frac{h}{\sqrt{2 \cdot m \cdot e \cdot V}}$ and amounts to $\approx 0.12\,\text{Å}$ for $10\,\text{kV}$ and $\approx 0.04\,\text{Å}$ for $100\,\text{kV}$ electrons. After a first approach by Ernst Ruska in 1931 using an electron beam for imaging in grazing incidence configuration [7], the first scanning electron microscope (SEM) was realized 1935 by Ernst Ruska and Max Knoll in Berlin [100]. 1965 the first commercially available SEM was distributed from Cambridge Instruments Company (U.K), followed six months after by a model from JEOL (Japan) [100].

In a SEM electrons are accelerated to typically $1 - 30\,\text{kV}$ and focused to a beam of spot size of $\lesssim 1\,\text{nm}$ with electrostatic and magnetic lenses. This beam is scanned along the sample surface, where it is backscattered and gives rise to secondary electron and X-ray emission. Different types of detectors are used to image different material parameters. Our Zeiss Auriga setup is equipped with two secondary electron detectors: one in Everhart-Thornley configuration [101] (called "SE2") and one inside the electron column ("in-lens"). Because the in-lens geometry is more sensible towards the sample work function, the in-lens detector provides generally a better material contrast than the SE2 [102]. Besides that, the AFM stage placed at our SEM chamber hinders a significant amount of the released secondary electrons to reach the SE2 detector, resulting in lower signal intensity. Therefore only SEM images gathered with the in-lens detector are presented in this work.

Our SEM is specified for a resolution of $1.2\,\text{nm}$, which is reached on material transitions with very high contrast (i.e. highly differing secondary electron emission) only. With the material systems analyzed here a resolution of $10 - 30\,\text{nm}$ is reached.

---

[7]In 1986, he was awarded with the nobel prize for this invention [63].

3.1.4 Focused ion beam microscopy

After Feynman proposed the use of ion beams for nanomanipulation in his groundbreaking lecture "There is plenty of room at the bottom" in 1959, the first microscopes based on hydrogen ions were realized in the mid-1970s from Escovitz et al. [103] and Orloff and Swanson et al. [104] in the U.S..

Focused ion beam (FIB) microscopy is based on the same principles as SEM, using accelerated ions instead of electrons. Usually secondary electrons are used for imaging, allowing for the application of identical detectors in SEM and FIB imaging. Because of its ability to sputter material with nm resolution, FIB is not restricted to imaging and material analysis, but allows also for nanofabrication (see Volkert and Minor et al. [105] and Notte [106] for good introductions in FIB technology).

For most FIB applications the ions are accelerated to energies of $5 - 30\,\mathrm{kV}$, so the same energy range as in SEM is used. However, because of their higher mass the ions carry much higher momenta: the momentum of a gallium ion is about 370 times larger than that of an electron of same kinetic energy. Thus, they transfer energy *and* momentum to the sample and sputter material from the surface. The ability of high energy ions to sputter surface atoms is described by the *sputter yield* Y, which is defined for a certain ion-target combination by

$$Y = \frac{\#\text{sputtered target atoms}}{\text{incident ion}}.$$

The first commercially available FIB systems were based on gallium ions. However, after the effective and ultrahigh resolution nanofabrication with helium [107] and neon [108] ions was proven in the late 2000s, FIB systems based on these ions were developed. Today, He and Ne FIBs are commercially available and broadly used for ultrahigh resolution nanofabrication and non-destructive imaging [109–111]. In this work, a *Zeiss Orion NanoFab* allowing for nanofabrication with He, Ne and Ga ions was used [112].

The determination of the sputter yield for a certain ion-target combination is not straightforward at all [113]. However, there are calculators available based on semi empirical formulas providing sputter yields that are in very good agreement with experimental values [114]. In this work, the Yamamura

3.1 Analytical methods

FIB ion	C_{12}	Si_{28}	Ag_{108}	In_{115}	Al_{27}	Ti_{48}	Au_{197}
He_4	0.02	0.02	0.2	0.2	0.05	0.02	0.1
Ne_{20}	0.4	1	4	4	1	1	2
Ga_{70}	2	2	17	16	5	3	18

Table 3.1: Sputter yield values $Y = Y$(sputter ion,target) for the ion/target combinations used in this work. The mass of the sputter ions/target atoms is given.

calculator from Arizona State University was used to calculate the sputter yields for He^+ and Ga^+ ions. The values for Ne^+ ions are taken from tables calculated by scientists of the NPL [115] in London with formulas derived from Seah [116]. To simulate sputtering of ITO, sputter yield values of its main component Indium were taken from [115] and [117]. In table 3.1 all values used within this work are listed. As can be seen, the sputter yield for a certain target material follows the trend $Y_{Ga} > Y_{Ne} > Y_{He}$ given by the order of the ion masses $m_{Ga} > m_{Ne} > m_{He}$. However, there is no intuitive access to the actual values, so experimental or semi empirical determination is inevitable.

At our Zeiss Auriga setup at iL, gallium ions are emitted from a liquid metal ion source. The ion beam current can be tuned from 5 pA up to 50 nA allowing for high resolution imaging in the low current range (with a resolution of about 10 nm) as well as fast nanomanipulation in the high current range.

3.1.5 BRR integrated SEM-AFM

All the above mentioned methods are available at the BRR integrated SEM-AFM [8] [118]. It consists of a Zeiss Auriga Crossbeam system and a scanning probe station from DME installed within the vacuum chamber of the crossbeam. This system allows for the FIB preparation and in-situ characterization of the prepared samples. The instrument has electric throughputs to the AFM stage, so that samples or electronic devices assisting the measurement can be addressed with a bias voltage during the characterization. The results

---
[8] "BRR" refers to the inventors of STM/AFM (Bining and Rohrer) and SEM (Ruska), emphasizing the integration of atomic force and electron microscopy in one system.

3 Experimental details

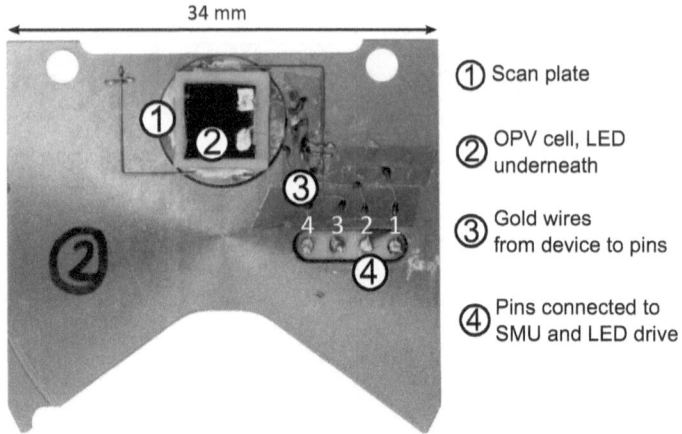

Figure 3.4: Sample holder for in-operando SKPM studies. The (small) solar cell (2) is illuminated by a LED located right below the sample and connected to an SMU via vacuum throughputs (4). To not interfere with the AFM/SKPM measurement, very thin gold wires (diameter= $0.05 - 0.1$ mm) are used for contacting (3).

presented in section 4.4 were obtained using sample holder equipped with electric connections for the organic solar cell under observation and a LED to illuminate it (see figure 3.4). With this, FIB prepared OSC samples are investigated in-operando at any working point along their charge transport path. As visible in figure 3.4, the sample is restricted to a maximum size of about $7 \cdot 7 \, \text{mm}^2$. For a more detailed description of the setup it is referred to chapter 4.5 of the dissertation of Rebecca Saive [119].

3.1.6 Transmission electron microscopy

Transmission electron microscopy (TEM) is a very powerful tool to study the structure-function relationship in OPV. In TEM a beam of monochromatic electrons propagates through an ultrathin sample where the electrons undergo interactions generating distinct patterns on a visualizing screen. There is a multitude of techniques based on the TEM configuration providing insights in different parameters of the sample, for example morphology, material composition and crystallinity [120]. In this work data from TEM diffraction and analytical TEM is presented, so their fundamentals are briefly discussed

3.1 Analytical methods

here. For a detailed description of TEM it is referred to the dissertations of Diana Nanova [37] or Martin Pfannmöller [121].

TEM diffraction

In TEM diffraction measurements the diffraction pattern of the monochromatic beam is used to study the crystalline order of the sample. Whereas highly ordered films lead to discrete diffraction spots, our amorphous or polycrystalline organic films lead to concentric rings. If the ordered domains are small compared to the thickness of the film and of arbitrary orientation, the sharpness of these rings is a measure for the degree of crystallinity. The distance of the rings from the center of the pattern is reciprocal to the spacings between the molecules in the film. For a quantitative analysis, radial profiles are taken and plotted in units of reciprocal nanometers [9].

Analytical TEM

Analytical TEM is a spectroscopic TEM approach that allows for real material contrast also in low contrast compounds as observed in this work.

The electrons that are propagating through the sample undergo inelastic scattering and the electron beam becomes polychromatic (see figure 3.5). Thus, with a spectrometer acting as electron prism the beam can be studied on the energy domain. We use two different methods based on this principle: *Electron energy loss spectroscopy* (*EELS*) and *electron spectroscopic imaging* (*ESI*). With *EELS* the whole electron spot is investigated concerning energy losses attributed to characteristic excitations in the sample. We use these characteristic excitations to determine the spatial distribution of a material compound. *EELS* spectra are extracted from samples with pristine materials to identify their characteristic plasmon peak energies. These characteristic energies we use in *ESI* series to interpret the gray scale values. In *ESI* (or *energy filtered TEM*) only electrons with a specific loss energy contribute to the image acquisition. With this, phase separation can be studied with a lateral

---

[9]The intensity of the diffraction pattern decreases inherently with increasing (reciprocal) distance from the center because of the increasing area the electrons are scattered to (increasing radius with constant solid angle magnitude). To account for this and preserve quantitative analysis, the presented profiles are obtained by concentric integration around the center of the pattern rather than by simple line profiling.

3 Experimental details

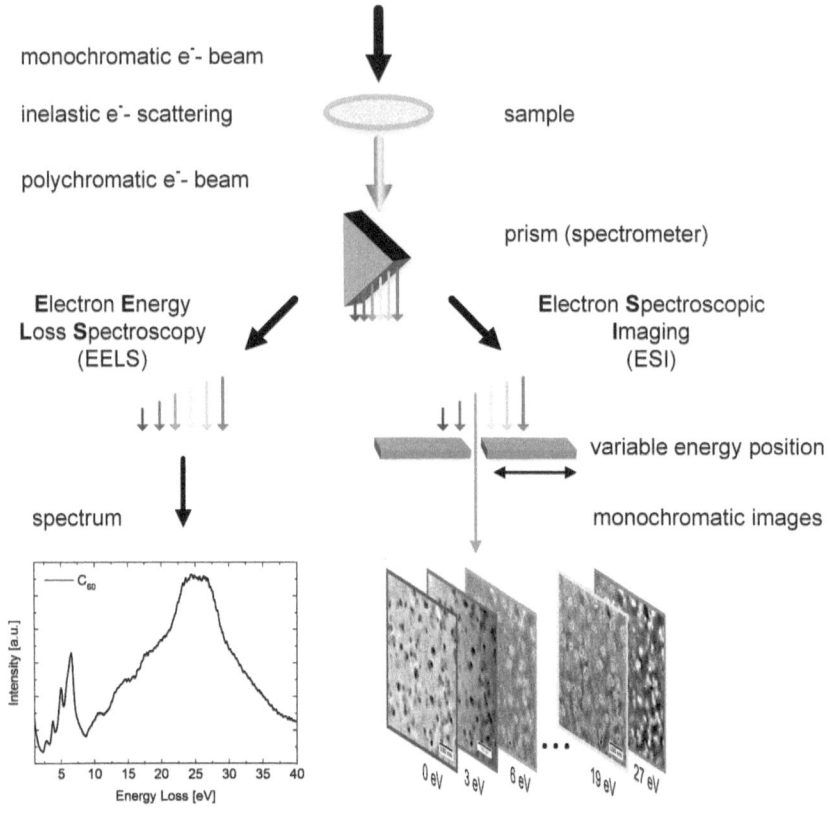

Figure 3.5: Principle of analytical TEM. The monochromatic beam becomes polychromatic after interaction with the sample, where electronic and plasmon excitations as well as ionization of core electrons gave rise to energy losses. The spectrometer separates the different loss energies, which are plotted spectroscopically (EELS imaging) and monochromatically (ESI imaging). Reprinted from [37].

3.1 Analytical methods

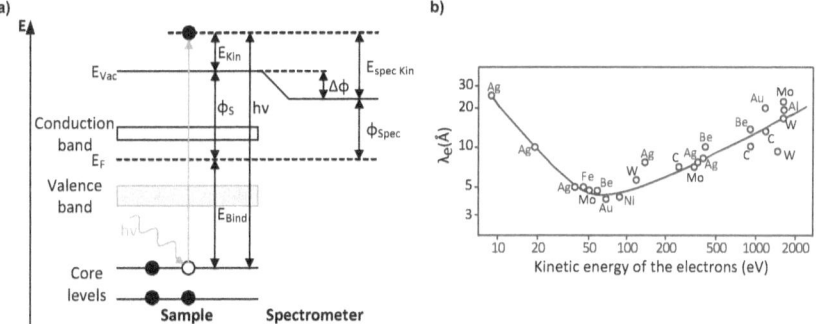

Figure 3.6: a) Working principle of photoelectron spectroscopy (PES). Electrons in the sample atoms absorb photons and are emitted through the sample surface. The transferred photon energy is consumed to overcome the binding energy E_{bind} and the work function difference $\Delta\phi$ between sample and spectrometer, excess energy turns into kinetic energy which is analyzed in the spectrometer. b) Inelastic mean free path of electrons in different solids. Reprinted from [77] (a) and [125] (b).

resolution down to a few nm. Recalling the crucial role of nanoscale phase separation in bulk heterojunctional OPV (see 2.2), analytical TEM is a key analytic tool in OPV research to fine-tune the BHJ morphology. In *plasmon peak mapping*, *ESI* data is evaluated with sophisticated software to achieve material class segmentation. Analytical TEM with plasmon peak mapping was successfully applied in solar cell research several times: Pfannmöller et al. could correlate the beneficial effect of annealing on the PCE to improved phase separation in (solution processed) BHJ OPV cells [122], Nanova et al. correlated the PCE of solution processed Perovskite solar cells to the degree of interconnectivity within the TiO_2 scaffold [123].

3.1.7 Photoelectron spectroscopy

From the methods available at the integrated UHV clustertool system, besides AFM also photoelectron spectroscopy (PES) was used in this work. Here it is briefly discussed, for a detailed description it is referred the dissertation of Julia Maibach [124].

For interface experiments we use X-ray photoelectron spectroscopy (XPS) to monitor the chemical properties of the sample and ultraviolet photoelectron spectroscopy (UPS) to study its work function and HOMO (or valence band)

3 Experimental details

onset [10]. The principle of PES is depicted in figure 3.6 a): photons with appropriate energy hit the surface of the sample and electrons are emitted according to the (outer) photoelectric effect [127]. Depending on the energy of the incident photons one distinguishes between XPS (using the Al K α line at 1486 eV and probing core level/inner shell properties) and UPS (using the He1 line with 21.22 eV and the He2 line with 40.8 eV and probing valence band/outer shell properties). Application of the conservation of energy brings

$$\hbar\omega = E_{bind} + E_{kin} + \phi_{sample} \implies E_{bind} = \hbar\omega - E_{kin} - \phi_{sample}. \quad (3.8)$$

This provides the sample work function with respect to the "vacuum level at infinity" [11]. Considering the work function difference of spectrometer (ϕ_{spec}) and sample leads to

$$E_{bind} = \hbar\omega - E_{kin} - \phi_{sample} + (\phi_{sample} - \phi_{spec}) \implies E_{bind} = \hbar\omega - E_{kin} - \phi_{spec}, \quad (3.9)$$

with the binding energy independent of the work function of the sample. The work function of the spectrometer is determined via a calibration measurement with a silver standard, so that the binding energy of the emitted electron can be calculated. Spectroscopic analysis of the emitted electrons (i.e. filtering of electrons with certain E_{kin}) yields information on band structure and work function (UPS) as well as presence and chemical bonds of certain elements (XPS). It is exploited that the core level energy position of an element constitutes a specific material feature. This position undergoes slight shifts which depend on the chemical surrounding (*chemical shift*). For the determination of the work function of the sample ϕ_{sample} with UPS one uses the emitted photoelectrons of zero kinetic energy E_{kin} and maximum binding energy E_{bind} (for the energetic configuration see figure 3.6 a)): this condition is met at the onset of the PES intensity, called the *secondary electron (SE) cutoff*. The electrons at the SE cutoff have right enough energy to overcome the work function and leave the sample remaining with no kinetic energy.

---

[10] For an introduction of PES in material science see Klein et al. [126].
[11] The reading of Cahen and Kahn [78] is highly recommended for a proper understanding of the concepts and parameters used here.

Equation 3.8 with the SE cutoff condition $E_{bind} = E_{bind}(SE)$, $E_{kin} = 0$ brings

$$\phi_{sample} = \hbar\omega - E_{bind}(SE).$$

The photoelectron analysis relies on elastically scattered electrons only, which limits the information depth of PES measurements. As depicted in figure 3.6, the mean free path λ_e of electrons in matter follows a tub shape with $\lambda_e < 1\,\mathrm{nm}$ for all energies relevant here. Considering UPS, there is an information depth (given by $3\lambda_e$) of only $\approx 15\,\mathrm{\AA}$. This makes PES a very surface sensitive method, allowing for thickness measurements on ultrathin films. As most samples characterized here grow in an amorphous and homogeneous manner, a Lambert-Beer ansatz yields sufficiently good results for the PES intensity attenuation caused by the film. With the intensity of the substrate emission line without the film being I_0, the intensity I of the PES signal after deposition of a film of thickness d is given by

$$I = I_0\, e^{-d/\lambda} \tag{3.10}$$

with the mean free path λ_e.

The energy resolution of PES amounts to about $100\,\mathrm{meV}$ for XPS and $50\,\mathrm{meV}$ for UPS.

3.2 Vacuum preparation of small molecule solar cells

In this chapter some background about the device and sample preparation at clustertool is presented. Following the footsteps of Ilja Vladimirov [128], Benjamin Martini [129] and Paul Heimel [130], who vacuum processed organic light emitting diodes (OLEDs) at iL, we realized an OPV preparation chamber (OLED1 in figure 3.1). Especially the design and installation of a substrate heating option for temperature control during the evaporation process without renouncing on the substrate rotation posed a serious challenge in terms of UHV engineering. It was met not least because of very valuable advice from Sebastian Montzka (Braunschweig group at IHF) and excellent engineering

3 Experimental details

by Felix Schell (now at Max-Born-Institut Berlin). When applying analytics at clustertool, the vacuum preparation of small molecules OPV has several advantages over the previously employed solution processing of polymer materials. This especially, when it comes to

- structural and morphological investigations with ultrahigh resolution techniques, such as analytical TEM (see chapter 3.1.6). On the one hand, the parameter space in terms of intentional manipulations such as a) thickness variation, b) vertical concentration gradients using co-evaporation and c) film crystallinity [12] is much broader and easier to access compared to solution processing. On the other hand, the parameter space in terms of unintentional manipulations caused by variations of the processing environment is reduced.

- very small substrates as the ones required in the BRR SEM-AFM (see chapter 3.1.5). In spincoating, it is impossible to achieve homogeneous layers on such substrates because edge effects extend all over the substrate. In vacuum preparation, the substrate size is not relevant at all.

- the correlation of several characterization techniques at clustertool (see our studies on the hole extracting contact presented in section 4.2). By means of several series of layer deposition and subsequent measurement on the surface, the growth of the OPV stack can be studied in great detail. Furthermore, the very high chemical sensitivity of XPS can be exploited to a much higher extent when working with materials that underwent sublimation purification in UHV.

Here, in addition to the facilities for vacuum preparation the materials and device stacks used in this work are discussed.

---

[12] In principle, substrate heating during evaporation offers the same additional thermal energy to all molecules within the organic film. However, the effect of enhanced diffusivity is much stronger on the (arriving) molecules at the very surface than on the ones already embedded in the bulk matrix. Therefore, vacuum preparation under substrate temperature control allows for the realization of vertical crystallinity gradients within an organic thin film. Obviously, the same applies for substrate cooling.

3.2 Vacuum preparation of small molecule solar cells

3.2.1 Materials and stacks

The active material molecules studied in this thesis are F_4ZnPc and C_{60}. Their chemical structure and energy levels are depicted in figure 3.7. However, for the realization of efficient OPV some additional molecules have to be used. The energy levels of the most prominent materials used for solar cell fabrication are depicted in figure 3.8, the OPV stacks in use are shown in 3.9. In the following, the most relevant properties of these materials and the parameter used for deposition are discussed.

Because we are primarily interested in the optical properties of the materials here, we quote the optical (excitonic) band gap of the materials (see the discussion on organic semiconductor band gaps in 2.1.2).

Fluorinated Zinc Phtalocyanine F_4ZnPc

F_4ZnPc is a partly fluorinated derivative of the planar molecule ZnPc, where four of the hydrogen atoms of ZnPc are replaced by fluorine (see figure 3.7). ZnPc as well as other phtalocyanines with a central metal atom (especially CuPc) are widely used as donor materials and were of the pioneer molecules in vacuum processed OPV: CuPc was used by Tang [8] to realize the first bilayer OPV exceeding 1% PCE, first studies on the photovoltaic and rectifying effects of phtalocyanines were published by Putseiko already in 1948 [131]. The fluorination at F_4ZnPc results in a stronger binding of the valence electrons and thus in a lowering of both HOMO and LUMO energy level of about 400 mV compared to ZnPc [17], without affecting its optical band gap and the excellent absorption properties. However, the lowering of the HOMO level leads to a higher effective band gap (see figure 3.7 c)), what allows for a higher extractable V_{oc}. Meiss et al. [17] achieved an increase in V_{oc} from 520 mV to 680 mV replacing ZnPc through F_4ZnPc.

Although p-conducting, F_4ZnPc as well as ZnPc have rather poor hole mobilities. Lacking of reliable results for Zinc Phtalocyanines, we refer here to the zero-field mobility values of pristine CuPc thin films determined by Opitz et al., who found $\mu_h = 1.8 \cdot 10^{-3}\,\mathrm{cm^2(Vs)^{-1}}$ in an organic field effect transistor geometry [132] and $\mu_h \approx 10^{-5}\,\mathrm{cm^2(Vs)^{-1}}$ in a SCLC diode geometry with organic layers of 200 nm thickness [133]. Several studies are available facing growth [133, 134] and electronic structure [135] of the

3 Experimental details

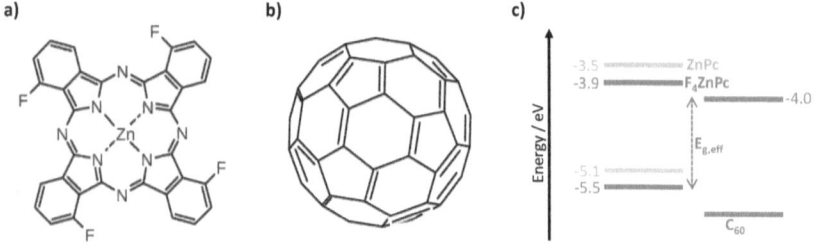

Figure 3.7: Chemical structures of a) F_4ZnPc and b) C_{60}. c) The fluorination of ZnPc leads to a higher effective band gap in the material combination F_4ZnPc/C_{60}.

mentioned Phtalocyanines. Schünemann et al. demonstrated that in thin films the α- and γ-phases dominate [136]. The F_4ZnPc used in this work was provided by BASF SE, where it underwent a purification by vacuum sublimation. By means of keeping the material constantly at $\approx 180°C$ at our evaporation chamber, it was further purified.

Buckminster fullerene C_{60}

The chemical structure and energy levels of C_{60} are shown in figure 3.7. C_{60} is the most prominent acceptor molecule in OPV because of its very strong electron accepting character[Savoie2014] and its high as well as isotropic electron mobility. Electron mobilities as high as $\mu_e = 5\,cm^2(Vs)^{-1}$ were reported for highly crystalline C_{60} thin films grown on templating substrates [137], Opitz et al. determined the zero-field mobility of a pristine C_{60} layer in a 200 nm SCLC diode geometry to be $\mu_e \approx 10^{-1}\,cm^2(Vs)^{-1}$ [133]. Still, this is 4 orders of magnitude higher than the hole mobility of CuPc in the same configuration. Growth studies were performed by Faiman et al. [138]. C_{60} with a purity of $> 99.9\%$ was purchased from ADS and further purified by keeping it constantly at $\approx 220°C$ at our evaporation chamber.

1, 3, 5-Tris(phenyl-2-benzimidazole)benzene TPBi

TPBi is used as buffer layer between C_{60} and the silver top contact in the non-inverted device configuration (see figure 3.9). It is transparent, has a high electron mobility, excellent hole/exciton blocking properties [139, 140]

3.2 Vacuum preparation of small molecule solar cells

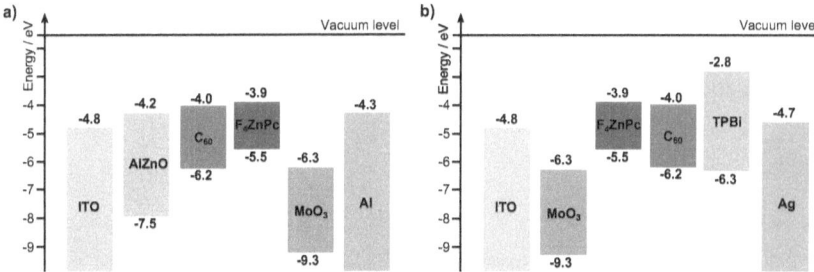

Figure 3.8: Optical band gaps of the materials used here in a) inverted and b) conventional solar cells. The extraction and injection layers AlZnO, MoO$_3$ and TPBi have large optical band gaps, so that losses caused by parasitic absorption are low.

and aligns energetically very well with the electron acceptor C$_{60}$ (see figure 3.8). Besides that, it reduces the roughness of the active layer, what prevents leakage paths through the device and preserves a high V_{oc} (see our results in chapter 5.1.3 and [141]). TPBi was purchased from Sensient Technologies.

Molybdenum oxide MoO$_3$

MoO$_3$ is used as a transparent hole extracting interlayer in the non-inverted as well as the inverted devices (see figure 3.9). It is known to effectively extract/inject holes in organic semiconductor devices [142] and increase their lifetime significantly [143]. The beneficial effect of the increased anode work function is somewhat foiled by its poor electron blocking properties [144]. At the iL analytics group, there is a broad experience using MoO$_3$ for applications in organic electronics (see the theses of Daniela Donhauser [145], Tobias Glaser [146], Sebastian Beck [147] and Maybritt Kühn [148]).

Aluminum doped zinc oxide AlZnO

Al doped ZnO is a transparent interlayer widely used in organic electronics because of its small work function and good electron extracting/injecting properties. We used AlZnO for the modification of ITO substrates, making them efficient electron extracting contacts when used in combination with C$_{60}$

for inverted devices. The AlZnO coated ITO substrates have a work function of $4.4 \pm 0.2\,\text{eV}$. The AlZnO nanoparticle ink *Nanograde N-20X* was provided from the KIT device group at iL.

Nickel oxide NiO_x

NiO_x is a transparent electron blocking layer used in organic [144, 149] and Perovskite [150] solar cells because of its excellent electron blocking properties. Here, we used NiO_x for the modification of ITO substrates, making them efficient hole extracting contacts when used in combination with F_4ZnPc in conventional (non-inverted) OPV devices. The NiO_x coated ITO substrates have a work function of $5.0 \pm 0.2\,\text{eV}$. The NiO_x precursor "nickel acetate tetrahydrate" was purchased from Sigma Aldrich.

Indium tin oxide ITO

ITO is the mixture of the metal oxides In_2O_3 and SnO_2. As in many applications like smartphone displays, ITO coated glass is the "backbone" of our solar cell stack (see figure 3.9). Because of its work function of about $4.8\,\text{eV}$ (which can be elevated to $5.4\,\text{eV}$ by O_2 plasma treatment), in OPV it is typically used as hole extracting contact. Nevertheless, it can be turned into the electron extracting contact by tuning its work function to values as low as $3.2\,\text{eV}$ [151]. Although less interesting for low cost large area OPV because of its parasitic absorption in the visible range and its relatively high price, its very low sheet resistance, the high chemical stability and its commercial availability make it still the "workhorse" in most OPV labs. Chemically it contains 74% In, 18% O_2 and 8% Sn.

3.2.2 Experimentals- deposition parameter

We prepared devices at the clustertool preparation chambers using structured ITO coated glass substrates of $25 \cdot 25\,\text{mm}^2$ with an active area of $4\,\text{mm}^2$ (BASF "OLED" layout). The substrates were cleaned in a ultrasonic bath for 10 min in acetone and isopropanol each, followed by 5 min of O_2 plasma treatment. For a more detailed description of the setup and processes involved see the master thesis of Felix Schell [152].

3.2 Vacuum preparation of small molecule solar cells

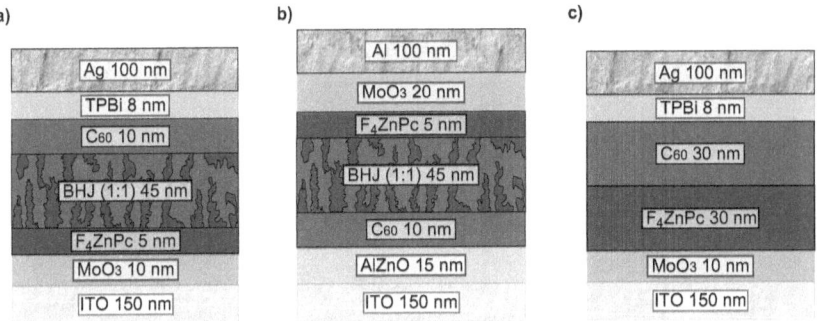

Figure 3.9: Organic solar cell stacks used in this work for BHJ solar cells in a) conventional and b) inverted architecture. c) Bilayer solar cell stack in conventional architecture.

If not mentioned otherwise, for all samples presented the organic materials F$_4$ZnPc, C$_{60}$ and TPBi were applied using a deposition rate of $10 \pm 2 \, \frac{\text{Å}}{\text{min}}$. MoO$_3$ was applied with deposition a rate of $15 \pm 5 \, \frac{\text{Å}}{\text{min}}$, silver and aluminum with a deposition rate of $90 \pm 20 \, \frac{\text{Å}}{\text{min}}$.

NiO$_x$ was applied via spincoating as described from Manders et al. [149]. We added 25 mg of monoethanolamine (from Sigma Aldrich) and 100 mg nickel acetate tetrahydrate to 4 ml of ethanol. After stirring it for 4 hours at 70° Celsius, the solution was spincoated 60 s with 4000 rpm and 2000 rpm/s. The substrates were post-annealed at 275 °C for 5 min at ambient air. This procedure yields closed NiO$_x$ layers with a thickness of 15 ± 5 nm.

AlZnO was applied via spincoating. The nanoparticle ink "Nanograde N-20X" from "nanograde" was diluted with 1-Propanol in the ratio 1:3. The solution was stirred for at least one hour at 50 °C and spincoated 60 s with 2000 rpm and 2000 rpm/s. The substrates were post-annealed at 120 °C for 10 min. This procedure yields closed AlZnO layers with a thickness of 20 ± 5 nm.

3.2.3 Vacuum thin film deposition

All devices presented here were prepared at the clustertool preparation chambers (see figure 3.1). The application of the thin films was performed by thermal evaporation of the materials and monitored with quartz microbal-

3 Experimental details

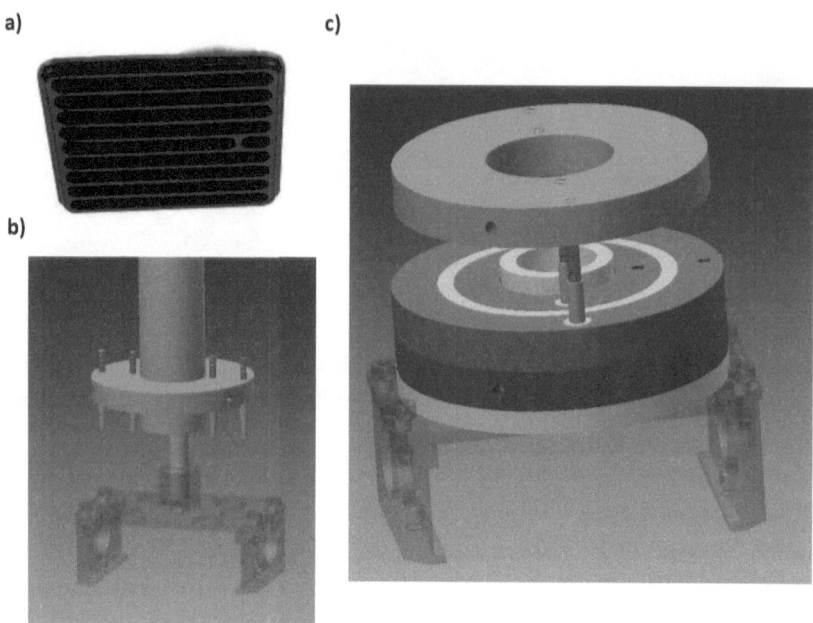

Figure 3.10: Substrate heating unit realized in the OLED1 organic preparation chamber. a) IR heating unit. b) Sample mount. The orange part carrying the sample holder rotates during deposition. c) Brass/copper sliding contacts. The upper (brass) part is at rest, while the copper part rotates with the sample mount. The contacts are isolated from the environment by (the white) Teflon parts.

ances. The resistive heaters surrounding the crucibles heat the material inside right above its sublimation temperature, resulting in a directed molecular beam moving towards the substrate. The microbalances were calibrated by thickness measurements performed with ellipsometry (SENpro, SENTECH Instruments). Each chamber is equipped with two microbalances placed vis-a-vis, allowing for the independent control over two sources in the same chamber. Thus 2-material-compounds with tunable mixing ratios can be applied by co-evaporation.

The crucial impact of the active layer morphology on the PCE in OPV necessitated the installation of a substrate temperature control unit. Because of the geometry of evaporation units and microbalances in OLED1, homogeneous thin films on $25 \cdot 25\,\text{mm}^2$ substrates can be grown under permanent substrate

3.2 Vacuum preparation of small molecule solar cells

rotation only. Thus, the strongest requirement on the heating system was the preservation of substrate rotation. In principle, this requirement could be met by a heating system consisting of a UHV capable heater placed in the cage of the sample mount. Thereby, the heater has to be power supplied by sliding contacts that do not harm the UHV (!). After some investigations and great support from Sebastian Montzka, Olaf Flechtner and engineers from Elstein, we found a solution using (i) the UHV capable infrared heater *Elstein SFH/4* [13] and (ii) a sliding contact using the material combination of copper and brass which is known for being almost abrasion-free. The design of Felix Schell was realized in OLED1 and is shown in figure 3.10 and is based on the mentioned infrared heater and the copper/brass sliding contact. The two power supplying electrodes are separated from each other and from the chamber environment by Teflon parts. For power control we use a Eurotherm 7100A thyristor unit. Because the system does not allow for further cabling, we renounced on an online temperature control. Instead, the system was calibrated for certain temperatures and run with heating routines while the evaporation process.

3.2.4 Solar cell characterization

The presented solar cells were characterized under AM1.5 illumination provided by a solar simulator from LOT-QuantumDesign. The $I-V$-characteristics were recorded with a Keithley 2601B SYSTEM SMU accessed with the commercial software ReRa TracerIV-curve and the (way more convenient) in-house-built IV CURVES software coded by Alexander Müller-Brand [14]. The solar simulator setup was located in a nitrogen glovebox till April 2015, and moved to ambient afterwards. Therefore, the results presented in this thesis originate from two different characterization routines. The ones characterized before april 2015 were transferred from the clustertool glovebox to the characterization glovebox under nitrogen atmosphere. The samples characterized after the move to ambient were encapsulated in the clustertool glovebox with barrier foil from 3M and characterized at ambient.

---

[13] The great Elstein SFH/4 does, if outgased properly, not harm the UHV of $< 10^{-7}$ mbar at all. Its UHV mission lasts, so far, already more than two years. It never caused any complications. The acquisition costs amounted to 22.30 € (incl. VAT, excl. shipping)!
[14] For details to the IV CURVES software see Alexanders master thesis [153].

4 Electric potential distribution of F_4ZnPc/C_{60} small molecule organic solar cells

In this chapter studies on the electric potential distribution within F_4ZnPc : C_{60} organic solar cells are presented. Different characterization techniques employed on different measurement geometries are applied and their results compared. First, we determine the potential evolution of an entire OPV bilayer stack from in-situ KP studies under UHV conditions. These results are backed by a detailed XPS/UPS study of the hole extracting contact. Ion implantation in the organic matrix and its possible influence on cross-sectional SKPM results are discussed. We compare the behavior of helium, neon and gallium FIB ions in a simulation study and characterize OPV cross sections prepared by the different FIB ions with SKPM. Finally, in-operando SKPM measurements on the cross sections of bilayer cells with extended film thicknesses are presented and discussed in the light of the afore mentioned results.

The author thanks Sebastian Hietzschold for using his equipment and for the great experimental support during the KP measurement series presented in the following section.

The author thanks Patrick Reiser for performing the PES measurements and for the great support with the evaluation. Thanks also to Eric Mankel for valuable discussions on the data.

4.1 Electric potential distribution of the OPV stack under short circuit conditions

The electric potential distribution within an electronic device is the key measure for the fundamental understanding of its device physics. When it comes to studies on the electric potential distribution of entire organic electronic devices or on certain device interfaces, the most popular characterization method is a work function measurement series on the evolving device stack/interface [1]. The extracted band diagrams give a quick and instructive insight into the band alignment of the device. For this, the deposition process is interrupted at a specific vertical stack position, a work function measurement is performed and the deposition process is continued. In such a layer-by-layer manner the work function evolution along the whole stack/interface of interest is measured. In combination with some knowledge on the materials (band gap, HOMO and LUMO positions), an energy band diagram describing the equilibrium energetics of the whole stack/interface can be derived. Although UPS [154] is probably the most popular tool for such stack characterizations (and is used in a similar experiment presented in section 4.2), here we use a conductive SKPM cantilever to determine the work function in a Kelvin probe configuration. With this we can ensure high comparability to the cross-sectional SKPM measurements presented in section 4.4. Additional methods to map the electric potential distribution across a device are the characterization of delaminated layers [155] or the insertion of additional electrodes at different stack positions [156, 157].

Experimental details

Here the layer-by-layer technique is applied to extract the equilibrium electron band diagram of the $F_4ZnPc:C_{60}$ bilayer stack displayed in figure 3.9 c). For this, organic and metal thin films are prepared in the OLED1 and metal preparation chambers. The samples are transferred through the clustertool system to the UHV-SPM for characterization (for the clustertool architecture see figure 3.1). In the UHV-SPM chamber the conductive SKPM cantilever is used as Kelvin probe for the measurement of the CPD. After determining

---
[1] The defined growth of a stack/interface as described here is possible for vacuum processed samples only.

4.1 Electric potential distribution of the OPV stack under short circuit conditions

Work function	MoO$_3$	F$_4$ZnPc	C$_{60}$	TPBi	Ag
ϕ_{meas} (eV)	6.0	4.7	4.9	4.1	3.4
ϕ_{Lit} (eV)	6.9 – 5.5	4.4 – 4.7	4.5 – 4.9	4.0	4.1 – 4.8

Table 4.1: KP work function results from figure 4.1 a) (first row) in comparison with literature values (second row). We estimate an error of ±0.1 eV for our values. The literature values for MoO$_3$ were determined by UPS and KP, all other values were determined by UPS. Origin of the literature values for MoO$_3$: [158, 159], F$_4$ZnPc, C$_{60}$: [33, 160], TPBi: [139], Ag: [161].

their CPD, the samples are transferred back to the preparation chambers where the next layer is applied. In best case, this sample transfer takes about half an hour (one-way). Here, in a two-day experiment KP measurements on seven stack positions were taken. The samples were kept under UHV conditions during the whole preparation and characterization process. The pressure in the preparation chambers was $\leq 5 \cdot 10^{-7}$ mbar, the pressure in the transfer as well as in the characterization chambers was $\leq 2 \cdot 10^{-8}$ mbar. Before and after the measurement series, the cantilever was calibrated by means of an HOPG sample. With this we could assure meaningful CPD results and calculate the absolute work function values from the measured CPD.

One issue in these measurements is the contacting structure of the device stack. As usual for KP measurements the ITO substrate is connected via the sample holder to a common ground potential with the KP setup. Because of the UHV in-situ nature of the experiment, the following layers can not be connected directly to the cable used to connect the ITO. However, we want to emphasize here that there is always a direct electric connection between the topmost layer and the common ground. This is ensured by the laminar deposition extending all over the sample holder and the transfer process with pliers smearing the different layers of the stack.

Experimental results

The work function values determined from the growing OPV stack are displayed in figure 4.1 a) and table 4.1. Data from a (less detailed) additional

4 *Electric potential distribution of* F_4ZnPc/C_{60} *small molecule organic solar ce*

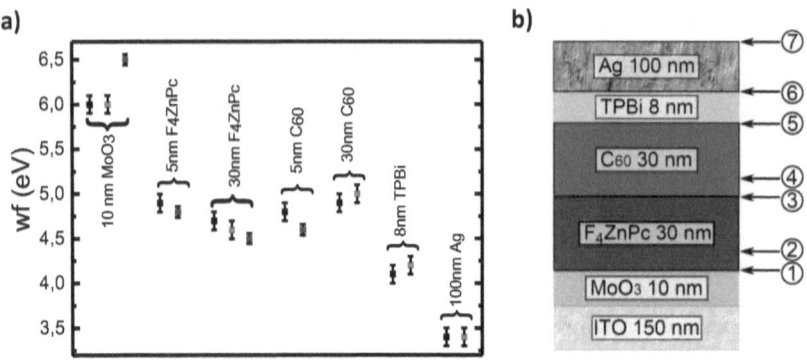

Figure 4.1: a) Work functions from layer-by-layer studies on the evolving OPV stack. b) The positions of the measurements are marked in the bilayer stack. KP measurements were taken after the deposition of: the MoO_3 layer, 5 nm and 30 nm F_4ZnPc, 5 nm and 30 nm C_{60}, TPBi and the Ag layer (black dots). For comparison, KP measurements on a second (control) stack are shown in red, UPS work function measurements on a sample prepared and measured in parallel are shown in blue.

KP measurement series performed on an identical sample and using a different cantilever are added. Also UPS data from the same sample are shown. Comparing the CPD data obtained from the two different KP measurements (black vs. red), we find a very good agreement over the entire data range. Comparing the KP and UPS work function values (black vs. blue), we find a very good agreement for the measurements on the organic layers (UPS values are generally lower than KP values [2]), but a significant deviation on the MoO_3 coated ITO.

In table 4.1 the values of our measurements are compared to values reported in literature. We find a very good agreement for the organic layers, but some deviations for the electrodes. MoO_3 is known for its fast work function decline even under UHV conditions [158]. However, relying on KP measurements on the time evolution of ϕ_{MoO3} in UHV from Hamwi [159], we would expect a value of $\phi_{MoO3} = 6.5$ eV rather than $\phi_{MoO3} = 6.0$ eV. This indeed resulted

---

[2] This is a method-inherent behavior and caused by the fact that in UPS, the very first (i.e. lowest bound) electrons leaving the surface of the illuminated sample area define the work function. In KP measurements however, the CPD value is obtained by averaging over the whole area influencing the Kelvin probe. Therefore, values obtained by UPS are generally lower than KP values.

4.1 Electric potential distribution of the OPV stack under short circuit conditions

from the UPS sample characterized within the same ±10 min. Yet, the reproducibility of the result in two independent measurements gives rise to the assumption that there is a method-inherent reason responsible for this result. The same applies to the reproducibly measured, very low value of $\phi_{Ag} = 3.4\,\text{eV}$. This reminds findings of Zhou et al. [61], who found effective work functions for metals in device stacks that were reduced by up to 0.8 eV with respect to their vacuum level values. They measured the built-in potential in MIM structures with OPV materials and found $\phi_{Ag} = 3.7 \pm 0.1\,\text{eV}$ for silver electrodes.

Nevertheless, for the overmost part of the data points we find a very good agreement between literature and our measurements with KP and UPS, highlighting the reliability of the data presented.

Extraction of the band diagram

From the measured work function values a band diagram displaying the electric potential distribution along the entire OPV stack under short circuit conditions is extracted. For this, additional information on band gap energies and HOMO levels/valence band maxima is required. Here, for F_4ZnPc and C_{60} we use UPS results from Corinna Hein [3] [33]. For MoO_3 values from Meyer et al. [142] are used, for TPBi values from Sun et al. [139]. Unfortunately reports on the electronic structure of TPBi are rare, so that the reliability of the values used is questionable. The resulting band diagram is shown in figure 4.2 a). Before discussing the findings, some remarks regarding the formation and the limits of this presentation are made (from left to right hand side in figure 4.2):

- The interface dipole δ between MoO_3 and F_4ZnPc is *estimated* to be $\delta = 1\,\text{eV}$. Relying on the work function as input parameter only, we have no means to separate the contributions of interface dipole and band bending. However, anticipating UPS results on the F_4ZnPc:C_{60} interface presented in the following section 4.2, we know that the UPS interface dipole is $\delta_{UPS} = 1.7\,\text{eV}$, the UPS band bending $eV_{bb}^{UPS} = 0.6\,\text{eV}$ and the overall UPS work function difference between MoO_3 and F_4ZnPc is 2.3 eV. Assuming the same ratio δ/eV_{bb} of dipole and band bending

---
[3] For the origin of the varying band gap values see chapter 2.1.2. Opposite to figure 3.8, here the PES instead of the optical band gap is used.

4 Electric potential distribution of **F_4ZnPc/C_{60}** *small molecule organic solar c*

in the KP and UPS measurements, we obtain $\delta_{KP} = 0.95\,\text{eV}$ and $eV_{bb}^{KP} = 0.35\,\text{eV}$ for the overall KP work function difference of $1.3\,\text{eV}$. This led to the values presented in the figure.

- We *assume* no interface dipole formation for all organic-organic interfaces. The F_4ZnPc/C_{60} interface was studied by Corinna Hein (see the band diagram in figure 4.2 b)), who found that no dipole is formed between F_4ZnPc and C_{60}. For the interface between C_{60} and TPBi, no information was available.

- Assuming $\delta = 0$, the band bending at the organic-organic interfaces is given by the difference in the work functions $\Delta\phi$. This band bending is always propagating into the following layer, i.e. considering the lack of information on the localization of the band bending, we *assume* it to propagate rightwards.

In figure 4.2 b) and c) reported band diagrams obtained from UPS studies on (parts of) the system characterized here are shown: Corinna Hein studied the $F_4ZnPc:C_{60}$ interface in detail (b) [33], Brendel et al. [160] determined the electronic structure of the cell containing the hole extracting contact with MoO_3 and the active (bi-) layer of the stack studied here (b). Comparing the diagrams we find a significant deviation between the band bending we estimated within the F_4ZnPc layer ($eV_{bb}^{F_4ZnPc} = 0.4\,\text{eV}$) to the values reported by Brendel et al. ($eV_{bb}^{F_4ZnPc} = 0.8\,\text{eV}$). Considering the work functions of the active layers, our values vary from the ones by Brendel et al., but fit the ones reported by Hein. Still, all studies see the same trend of a higher work function in the C_{60} layer.

In conclusion, despite the drawbacks discussed above and partly deviations regarding absolute values, we find that the electronic structure presented here gives a reliable picture of the band alignment within the bilayer cell. All electric potential gradients obtained here agree in the direction of the electric field with the ones reported.

Detailed electronic structure of the F_4ZnPc/C_{60} bilayer organic solar cell

The electron band diagram is presented in figure 4.2. Because transport properties are discussed we use PES band gaps here (see the discussion on

4.1 Electric potential distribution of the OPV stack under short circuit conditions

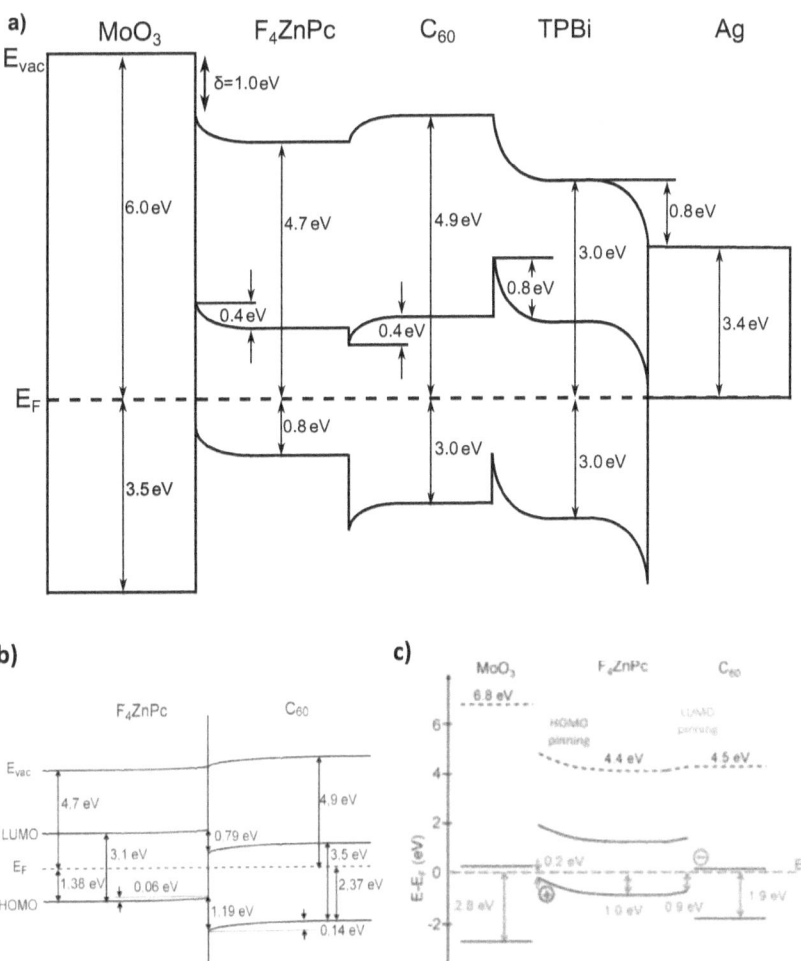

Figure 4.2: Electron band diagram with the results from our KP measurement (a). Band diagram from b) Corinna Hein and c) Brendel et. al. based on results from UPS studies. HOMO positions given by Hein are determined from the maxima of the HOMO resonance, Brendel et al. used the HOMO onsets. Because of this Hein got HOMO values that are higher by about 400 meV than the ones of Brendel.

organic semiconductor band gaps in 2.1.2). The electron band diagram gives detailed insights in the fundamentals of the studied bilayer OPV cells:

We find a positive bend bending towards the MoO_3/F_4ZnPc interface, leading to a drift field that supports the extraction of free holes from the F_4ZnPc HOMO level to the MoO_3 coated ITO contact. Also in UPS studies on MoO_3/AlPcCl [4] interfaces, this bend bending was found and identified as one of the main reasons for the beneficial effect of the MoO_3 anode interlayer in OPV.

Observing the F_4ZnPc/C_{60} DA interface we find a negative band bending in the C_{60}. This corresponds to a drift field hindering electrons to propagate towards the electron extracting contact. The disadvantageous band bending is a consequence of the fact that C_{60} has a higher work function than F_4ZnPc. Brought in contact, there is an electron flow towards the F_4ZnPc. This is also the case for many other DA material combinations used in OPV, for example $ZnPc/C_{60}$ and $CuPc/C_{60}$. Hein et al. demonstrated that this disadvantageous electric field at the DA interface can be reversed by p-doping of the donor material [162] or by a process protocol using elevated substrate temperatures [163]. Unfortunately their studies comprise UPS data only, leaving open if this field inversion indeed leads to elevated PCEs in OPV devices or is compensated by other effects.

The C_{60}/TPBi interface exhibits a large electron barrier of 1.1 eV. However, there are some reasons that advise not to over-interpret this result. Besides the very small database concerning the TPBi electronic structure, an interface dipole formation cannot be excluded here and could lower this barrier significantly. In addition, the working principle of such ultrathin buffer layers is still under debate. In an outstanding work Yoshida [164] demonstrated that the electron transport through a 10 nm thin BCP layer placed between C_{60} and the Ag cathode occurs via the LUMO level of a BCP-Ag complex located in the BCP band gap rather than via the BCP LUMO level. Therefore the energy barrier given here is questionable and we renounce a deeper discussion. Nevertheless, the band bending at the interface is beneficial for effective electron extraction. Same applies for the TPBi/Ag interface. In section 5.2.2 we present results on BHJ cells based on the same electron extracting contact

---

[4] Both AlPcCl and F_4ZnPc are high lying donor materials for OPV and comparable in terms of their HOMO positions.

studied here. Short circuit currents as high as $j_\mathrm{sc} \gtrsim 9\,\frac{\mathrm{mA}}{\mathrm{cm}^2}$ are achieved in these solar cells, underlining the quality of this electron extracting contact and questioning the existence of an electron barrier height of 1.1 eV.

Based on the band diagram we can comment on the built-in potential V_bi of the cell (for information on the concept of the built-in voltage in OPV see section 2.2.4). The difference in work function between the hole and the electron extracting contacts determined here is $\Delta\phi = 2.7\,\mathrm{eV}$. Taking into account the interface dipole at the MoO_3/F_4ZnPc interface, which does not contribute to V_bi, we find for the upper limit of the built-in potential: $V_\mathrm{bi} \leq 1.7\,\mathrm{eV}$. With electroabsorption spectroscopy (EA) on MIM diodes with a PEDOT:PSS anode, a silver cathode and different OPV materials, Zhou et al. found $V_\mathrm{bi} = 1.4 \pm 0.2\,\mathrm{eV}$ [61]. Also with EA, Siebert-Henze et al. found values $V_\mathrm{bi} \leq 0.9\,\mathrm{eV}$ that strongly depend on the used transport layers [59]. We underline here that the main contribution to this high estimation stems from the very low work function of the silver cathode.

Conclusion

From the layer-by-layer KP characterization of the entire F_4ZnPc/C_{60} OPV bilayer stack we got an integral picture of its internal electric potential distribution. We found a beneficial band bending in the F_4ZnPc donor layer supporting hole extraction. At the DA interface, we found a disadvantageous band bending, which hinders separated charge carriers on propagating towards the contacts. This band bending probably induces a significant bimolecular recombination at the DA interface, limiting V_oc. The upper limit of the built-in potential is $V_\mathrm{bi}^\mathrm{max} = 1.7\,\mathrm{eV}$.

4.2 UPS/XPS study on the hole extracting contact

Functional principle and performance of a charge extracting/injecting OPV contact depend critically on the electronic alignment and chemical processes occurring at the very interface. Detailed UPS/XPS studies covering the process of film growth from sub-monolayer to some tens of nm address these parameter with very high precision, allowing deep insights into the physics of contact formation. Here, the hole extracting interfaces consisting of the donor F_4ZnPc and two different anode materials were studied with UPS and XPS in a layer-by-layer manner. As high work function anodes, we used MoO_3 coated ITO and O_2 plasma treated ITO. After some details on the experiment we discuss the implications of the UPS results on the electronic structure of the contact. XPS data is consulted to comment on chemical interactions at the interface. The fundamentals of UPS/XPS are discussed in section 3.1.7.

Experimental details

Both substrates were cleaned in an ultrasonic bath of Acetone and Isopropanol for 10 min each, followed by a O_2 plasma treatment of 5 min. The samples were transferred to the UHV within 15 min after plasma treatment. The UPS measurement on the MoO_3 coated sample was performed $\approx$ 40 min after the MoO_3 evaporation, the measurement on the bare ITO was performed $\approx$ 35 min after plasma treatment. A non-plasma-treated ITO substrate was prepared using the same cleaning and measuring routine, renouncing the plasma step.

Analog to the KP experiment presented in the previous section 4.1, F_4ZnPc and MoO_3 thin films were prepared in the OLED1/metal preparation chamber and transferred to the UPS/XPS chamber for characterization. To apply thin F_4ZnPc layers reproducibly, deposition rates as low as $2\frac{\text{Å}}{\text{min}}$ were used for the smaller layer steps. In best case, this transfer takes about half an hour (one-way). In a two-day experiment, seven stack positions were characterized. During preparation and characterization, the samples were kept under UHV conditions (pressure of the preparation chambers $\leq 5 \cdot 10^{-7}$ mbar/characterization chamber $\leq 5 \cdot 10^{-9}$ mbar). Starting from submonolayer thicknesses (0, 2 and 8 Å) especially addressing the electronic structure of the substrate as well as chemical issues at the interface, the measurement series contain layers of 18, 36 and 90 Å tracing the band bending within the F_4ZnPc layer and are ending at 25 nm, where interfacial effects are decayed.

4.2 UPS/XPS study on the hole extracting contact

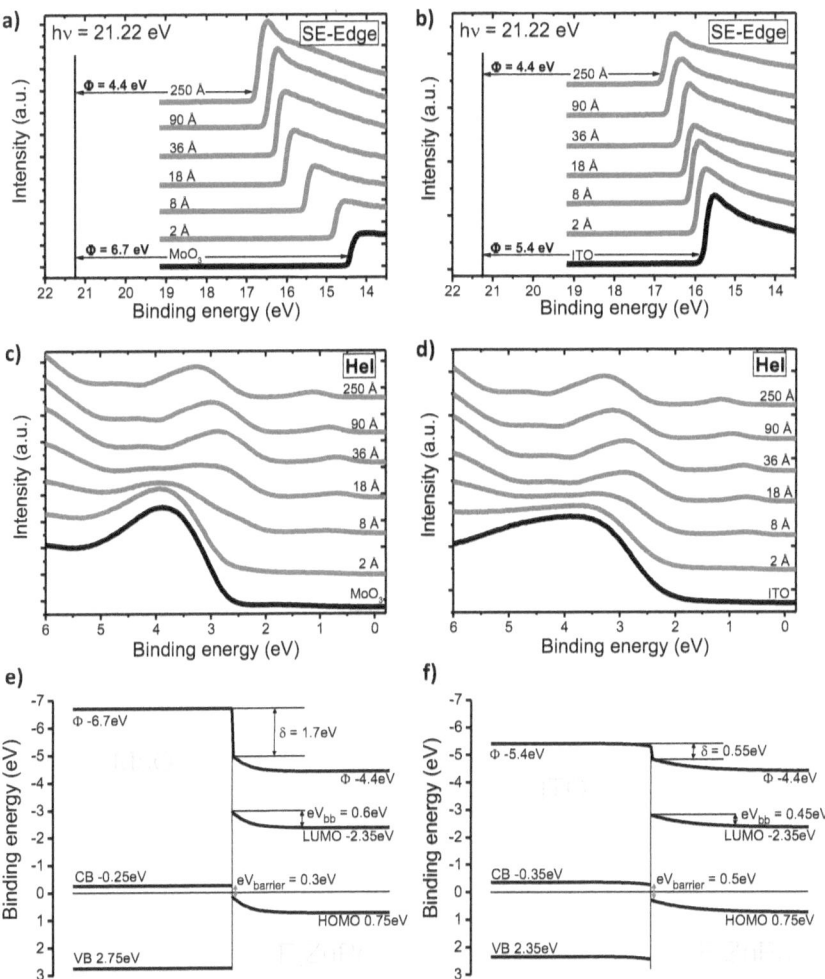

Figure 4.3: UPS data of F_4ZnPc grown on a), c), e) MoO_3 coated ITO and b), d), f) O_2 plasma treated ITO. a), b) From the secondary electron edge the work functions are extracted. c), d) From He1 spectra the HOMO onsets are determined. e), f) Electron band diagrams extracted from the UPS results. Larger band bending and lower hole extraction barrier are found for the MoO_3/F_4ZnPc interface.

4.2.1 Electronic properties

Figure 4.3 shows the UPS results for the different F_4ZnPc film thicknesses grown on both substrates as well as the band diagrams extracted from these data. The work function (wf) values are extracted by fitting the SE cutoff, the HOMO positions by fitting the HOMO resonance. We define the HOMO position from the *onset* of the respective He1 peaks.

MoO_3 coated ITO

For the MoO_3 substrate we found a wf of 6.7 eV, followed by a steep decrease leading to a strong interface dipole of $\delta_{MoO3} = 1.7$ eV (figure 4.3 a) and e)). Despite the existence of this strong dipole (consuming a large fraction of the total wf difference $\Delta\phi$), there is still a significant (upward) band bending of 0.6 eV (towards the hole extracting contact). To get a grasp on the impact of this band bending on charge transport in the donor layer, we compare the electric field resulting from this band bending to typical electric fields in OPV cells. Reminding the F_4ZnPc/C_{60} OPV bilayer stack displayed in figure 3.9 c) with a total (organic layer) thickness of $d_{stack} = 68$ nm and assuming linearity in the electric fields (as in Schottky depletion regions), the band bending of $eV_{bb} = 0.6$ eV extending over a band bending region of $d_{bb} \approx 15$ nm corresponds to an equivalent applied bias voltage of $V^{eq}_{applied} \approx 2.7$ V. This results in an electric field driving holes to the contact of a strength as high as $E_{bb} = 0.4$ MV/cm:

$$E_{bb} = \frac{V_{bb}}{d_{bb}} \approx 0.4\,\text{MV/cm} \approx \frac{V^{eq}_{applied}}{d_{stack}} = \frac{2.7\,\text{V}}{68\,\text{nm}}.$$

This shows that the band bending in the F_4ZnPc corresponds to an electric field more than 4 times higher than the (averaged) field induced by the actual driving voltage $V_{mpp} \sim 0.6$ V of the bilayer cell. Thus we assume that the band bending contributes significantly to an effective hole extraction.

We determined a hole extraction barrier of $\Delta_{holes} = 0.3$ eV. This is in good agreement with the value of $\Delta_{holes} = 0.2$ eV reported by Brendel et al. [160]. Same applies to the MoO_3 valence band maximum of 2.75 eV (Brendel et al.: 2.8 eV). Based on this detailed analysis of the hole extracting contact we can

4.2 UPS/XPS study on the hole extracting contact

modify the energy band diagram presented in figure 4.2. We use the data obtained here from the hole extracting contact region and the KP data from the former section 4.1. The refined energy band diagram is shown in figure 4.6 at the end of this section. Referring to the discussion on the built-in voltage in the previous section, based on the UPS data we correct the upper limit of the built-in voltage to $V_{bi}^{max} = 2.0 \, eV$.

Bare ITO

In case of the bare (O_2 plasma treated) ITO substrate we found a wf of 5.4 eV, an interface dipole of $\delta_{ITO} = 0.55 \, eV$ and a band bending in the F_4ZnPc layer of 0.45 eV (figure 4.3 b) and f)) . Following the assumptions discussed in the previous section, this corresponds to a built-in electric field of $E_{bb} = 0.3 \, MV/cm$ and an equivalent applied bias voltage of $V_{applied}^{eq} \approx 2.0 \, V$.

We find a hole extraction barrier of $\Delta_{holes} = 0.5 \, eV$, which is about 0.2 eV higher than for MoO_3.

Considering the He1 spectra shown in figure 4.3 d) and the band diagram in f), we find that the degeneration of the ITO [165], responsible for its high conductivity, is canceled by the plasma treatment. This is the case at least for the UPS information depth of $\approx 15 \, \text{Å}$, so for the very ITO surface (see section 3.1.7 for details on UPS). To have a closer look on this phenomenon, in figure 4.4 we present He1 spectra of a plasma-treated and a non-treated ITO sample. Right at the Fermi edge (0 eV binding energy) we find no feature in the signal of the plasma-treated ITO sample, whereas for the non-treated sample there is a significant signal increase. This shows that in the non-treated sample there is an occupation of electronic states at the Fermi level, proving the nature of ITO as degenerated semiconductor. After plasma treatment this is not the case anymore, the degeneration is canceled.

This leads to a reduced recombination velocity [166] at the contact, i.e. the ability of the contact for hole extraction is reduced: because less electrons are available right at the Fermi edge, the electron-hole recombination enabling hole extraction from the organic donor layer is suppressed (see Meyer et al. for the concept of hole extraction [142]). However, it is hard to tell if the observed effect is strong enough to substantially hinder hole extraction in our (decently current harvesting) bilayer devices, or if the remaining recombination velocity is still sufficiently high. Device data presented in section 5.1 indicate

4 Electric potential distribution of F_4ZnPc/C_{60} small molecule organic solar ce

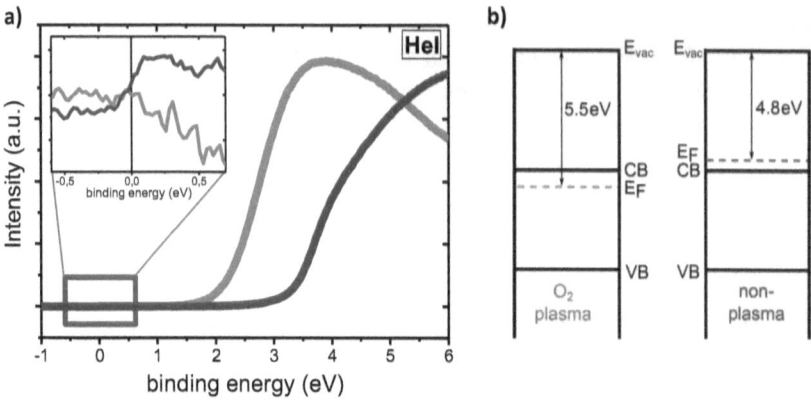

Figure 4.4: He1 spectra of O_2 plasma-treated (red) and non-treated (blue) ITO. a) The HOMO onset of the O_2 plasma-treated ITO (red) is shifted significantly towards the Fermi level when compared to non-treated ITO (blue). The zoom on the Fermi edge (0 eV binding energy) shows that for the non-treated sample there is a significant signal increase right at the Fermi edge, indicating the existence of occupied states at Fermi energy (degenerated semiconductor). The plasma-treated sample shows no signal increase at the Fermi edge. b) Electron band diagram of O_2 plasma-treated (left) and non-treated (right) ITO. The ITO degeneration is canceled for the plasma-treated sample, the Fermi level is located in the band gap.

the latter: FFs as high as 62% were achieved with O_2 plasma treated ITO substrates. A significantly undersized recombination velocity would lead to S-shaped solar cells with strongly reduced FFs [166].

4.2.2 Chemical properties

Figure 4.5 shows XPS data of the growing F_4ZnPc stack on both substrates. The measurements were performed simultaneously and on the same samples as the UPS measurements discussed above.

Before discussing the results we shortly address the impact of the work function evolution on the XPS data. In XPS the binding energy of core electrons is measured via its escape through the sample surface. So obviously, the binding energy values undergo shifting according to the evolution of the sample work function. When discussing chemical problems with XPS data in the following, we do not refer to these (continuous) shifts caused

4.2 UPS/XPS study on the hole extracting contact

by the evolving wf. XPS and UPS data were cross-checked to exclude misinterpretation of the XPS data.

The Mo3d and the In3d peaks document on the increasing attenuation of the substrate (MoO_3 and ITO) signals with increasing F_4ZnPc layer thickness. Regarding the F_4ZnPc growth (discussed more detailed in section 5.1.3), we observe a slightly stronger signal evolution of the F1s peaks on ITO than on MoO_3, indicating more efficient sticking of the F_4ZnPc to the ITO than to MoO_3.

O_2 plasma activates ITO

Very interesting is the differing F1s peak evolution on the MoO_3 and on the O_2 plasma treated ITO substrate, indicating strong chemical interaction between F_4ZnPc and the plasma-treated ITO. For a 2 Å F_4ZnPc layer on ITO we find the F1s peak strongly shifted by 2.7 ± 0.1 eV with respect to its actual position. With growing layer thickness this shifted peak decays and a second peak emerges at the actual F1s peak position. From peak analysis we find that the first 1/2 to 1 monolayer of the F_4ZnPc molecules undergo a chemical reaction with the ITO substrate, leading to less electronegative fluorine atoms. This seems odd in the framework of the F_4ZnPc molecule, since an additional negative charge is not expected to settle in proximity to the already electron enriched region of the Fluor atoms (see section 3.2.1 for details on the molecules). Based on literature reports we assume that a significant amount of the first F_4ZnPc monolayer decomposes when deposited on plasma-treated ITO. In-situ FTIR studies on the same system could unfortunately not further elucidate the observed phenomena.

Gassenbauer and Klein studied the growth of ZnPc on oxygen-enriched ITO in a layer-by-layer manner with XPS [167]. In a differing experimental setup (oxygen-enriched ITO bulk in UHV at their experiment versus oxygen-enriched ITO surface exposed to ambient here), they found a strong chemical interaction between ZnPc and ITO, in case that the latter was oxygen-enriched. They characterized sub-monolayer ZnPc coverages using synchrotron radiation, allowing signal-to noise ratios not achievable at our system. They found evidence for the creation of C-O complexes formed between oxygen from the ITO surface and C atoms released after the breaking of C-C and C-N bonds within the ZnPc molecule. In other words, they found that the

4 Electric potential distribution of F_4ZnPc/C_{60} small molecule organic solar ce

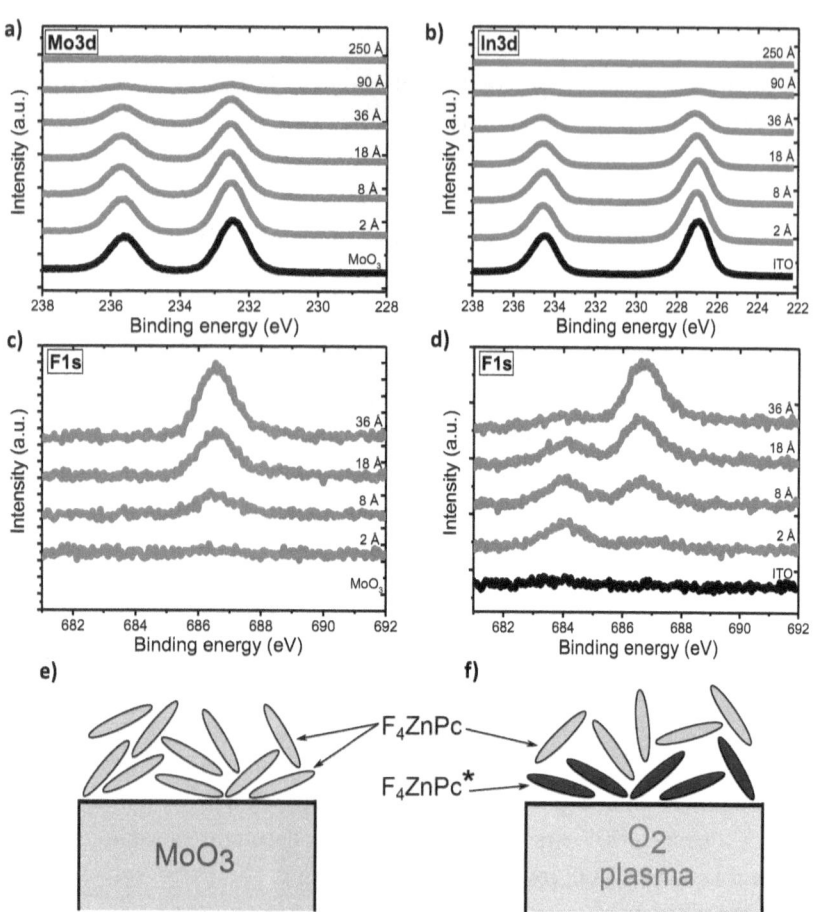

Figure 4.5: XPS data of F_4ZnPc grown on MoO_3 coated ITO (left) and O_2 plasma treated ITO (right). a), b) Mo3d and In3d peaks document the substrate signal attenuation with increasing F_4ZnPc layer thickness. c), d) F1s spectra from the first four evaporation steps. Whereas the F1s peak for F_4ZnPc grown on MoO_3 increases continuously at the same binding energy, there is a distinctive double peak evolution for F_4ZnPc grown on the oxygen-activated ITO. Parallel to the attenuation of the first peak with increasing F_4ZnPc layer thickness, a second peak emerges. The first peak is shifted by $2.7 \pm 0.1\,\mathrm{eV}$ with respect to the F1s peak of pristine F_4ZnPc. e), f) Sketch of F_4ZnPc growth on MoO_3 (e) and O_2 plasma-treated ITO (f) from a chemical point of view: in the latter case, a significant amount of the first monolayer consists of decomposed F_4ZnPc^*. Figures e), f) are adapted from [167].

oxygen activation of ITO can lead to a decomposition of organic molecules deposited onto the oxygen-activated surface. This strengthens our findings of the formation of a sub-monolayer consisting of decomposed F_4ZnPc right at the O_2 treated ITO/F_4ZnPc interface. As depicted in figure 4.5 f), intact F_4ZnPc growths on top of this decomposed layer.

On the MoO_3/F_4ZnPc interface we found no evidence for any chemical interactions: the deposited F_4ZnPc stays intact from the first monolayer on.

Conclusion

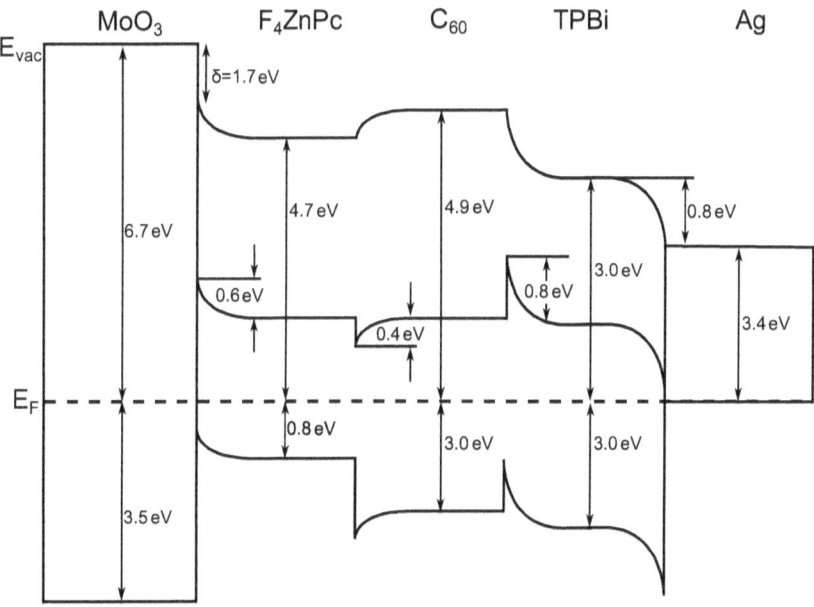

Figure 4.6: Refined band diagram of the OPV bilayer stack based (i) on the KP measurement series presented in the previous section and (ii) on the detailed UPS results from the MoO_3/F_4ZnPc interface presented here.

We presented a detailed comparison of the interfacial properties between a MoO_3/F_4ZnPc and an O_2 plasma treated ITO/F_4ZnPc OPV hole extracting contact. For both we found an energy alignment suitable for efficient extraction of free holes. This is in line with device data presented in section 5.1, which exhibit an increase of $\approx$ 30% in PCE when using MoO_3 instead of plasma treated ITO. As expected, the higher wf difference between MoO_3 and the F_4ZnPc leads to a stronger band bending than in the case of ITO ($eV_{bb} = 0.6\,eV$ versus $eV_{bb} = 0.45\,eV$). These band bendings lead to strong built-in electric fields as high as $E_{bb} = 0.4\,MV/cm$ and $E_{bb} = 0.3\,MV/cm$ in proximity of the contact, supporting hole transport in the poorly conducting F_4ZnPc. Also the hole extraction barrier is significantly smaller for MoO_3 ($\Delta_{holes} = 0.3\,eV$ versus $\Delta_{holes} = 0.5\,eV$). Exploiting the detailed dataset we could present a refined energy band diagram of the OPV bilayer stack. After plasma treatment the degeneration of ITO at the very surface is canceled. Still, from an energy point of view not only coating with MoO_3, but also O_2 plasma treatment allows for the establishment of decent contact formation between ITO and low lying donor materials such as F_4ZnPc. However, chemical analysis with XPS demonstrates that the beneficial effect of a good band alignment in the case of O_2 plasma treatment activates the ITO surface, leading to decomposition of a significant fraction of the first F_4ZnPc monolayer. This is not the case for MoO_3 coated ITO, where no chemical interaction at the interface was found.

4.3 The influence of FIB preparation on SKPM results

In this section we critically examine the FIB preparation of organic solar cell cross sections. The influence of FIB preparation on the electric potential distribution studied with cross-sectional scanning Kelvin probe microscopy (SKPM) is discussed in the light of simulations on FIB ion implantation in the device cross section. We compare FIB preparation using helium, neon and gallium FIB ions. We simulate the ion implantation into the device cross section for the different ion species and find inhomogeneous implantation along the cross section. The results are correlated with SKPM studies on samples prepared by FIBs with He, Ne and Ga ions.

The author thanks Dr. Anna Heidt and Dr. Endre Majorovits from Carl Zeiss AG for the FIB preparation of solar cell cross sections using He and Ne ions in the Zeiss Orion ion microscope [112].

4.3 The influence of FIB preparation on SKPM results

4.3.1 State of the art

The issue of exposing solar cell cross sections for studies along the main charge transport path of solar cells was addressed already in the 1970s, when electron beam induced current (EBIC) studies became a popular tool in solar cell research [168, 169]. Many of the inorganic solar cells characterized back then were composed of crystalline or polycrystalline materials. Here, the exposure of the cross section leads to a breakup of crystal bonds: the emerging dangling bonds are highly reactive and cause surface recombination and Fermi level pinning. The induced changes in local device performance and local potential distribution lead to strong surface recombination in EBIC [169] and (typically) built-in potentials smaller than bulk values in SKPM [170, 171]. In organic solar cells there is neither a long-range order nor are there covalent bonds (between molecules) that break up when the cross section is exposed. However, the active materials are very sensitive towards the exposure to air or the implantation of impurity atoms in the organic matrix. Saive et al. demonstrated that exposure of certain regions of an organic transistor to Ga FIB ions induce channels with conductivities several orders of magnitude higher than conductivities of the non-irradiated organic bulk [172]. SKPM studies revealed that the work function at these FIB exposed channels was increased by about 150 mV with respect to untreated sample areas, indicating p-doping through Ga ions.

On this basis we studied the influence of different techniques for cross section exposure with respect to their influence on cross-sectional SKPM results [173]: organic P3HT:PCBM BHJ solar cells were cleaved, microtome cut and FIB prepared and the resulting potential distributions compared. We found that samples prepared by all techniques lead to very similar SKPM results, but both reproducibility and cross section smoothness proofed best when Ga FIB was used. Based on these result we recommended on FIB milling as the method of choice. However, some doubts regarding this method remained. For example, all 0 V potential profiles of organic electronic devices indifferent of the particular material composition reported by Rebecca Saive exhibited a bowl-like shape [12, 13, 174]. This shape corresponds to a work function of the organic active layers that is higher by some 100 meV than the electrode work functions, which is not meaningful for all of the studied material combinations. Same applies to joint studies performed by Tobias

Jenne and the author on CBP/MoO$_3$ bilayer devices [175]: also here the CBP organic layer exhibited a higher work function than MoO$_3$, although the opposite is expected.

The effect of ion implantation by means of bombarding (inorganic) semiconductors with energetic ions (as here in the 10 kV regime) was extensively studied already in the 1950s and '60s [176] on the road to realization of high quality p-n junctions. A wide range of implantation atoms was studied from Cussins et al.. They found the p-doping effect observed indifferent of the particular electronic configuration of implanted atoms to be caused by sample damage rather than actual doping [177]. Annealing procedures were developed that ruled out beam damage as the dominant process parameter. With this it was demonstrated that p/n-doping could indeed be induced in a controlled way choosing appropriate III.-V. elements [178] or for example cesium for n-doping [179]. Studies addressing electric modifications induced by Ga FIB ion implantation were conducted on GaAs wafers [180] and MOSFETs [181]. As in Saives paper it was found that p-doping was induced by Ga FIB treatment. Very recent publications addressing the issue of electric modifications through He and Ne FIB treatment found that these FIBs induce n-doping, at least in the observed material systems. He FIB was used to nanopattern graphene [182] and MoS$_2$ layers [183] and in both systems n-doping was observed. Same was found when graphene was treated with Ne FIB [184]. In these studies the doping effect was primarily attributed to structural modifications on the surface. For example, in the graphene samples n-doping was induced by the replacement of carbon by nitrogen atoms [182]. However, these examples does not rule out that in other materials also p-doping could be induced. Unfortunately, in non of these studies different FIB techniques were compared directly in a measurement series on identical samples as was done here.

On this background the following study was performed. Exploiting the new FIB techniques using He and Ne ions, we aimed to learn about the FIB influence on SKPM results by comparing measurements on cross sections of identical solar cell samples exposed by different FIB ions. Of course, FIB preparation always causes the implantation of ions in the cross sections. Our reasoning was that still, ions of the electrically less active noble gases He and Ne implanted in the organics would be less invasive concerning the electric

4.3 The influence of FIB preparation on SKPM results

properties of the cross section and exploiting this we could shed some light on the preparation of organic device cross sections with FIB.

4.3.2 Experimental details

In the following SKPM data gathered from organic solar cell cross sections exposed by different FIB ions are evaluated. We prepared F_4ZnPc/C_{60} bilayer solar cells with thick active layers of 300 nm each according to the standard procedures described in section 3.2. Besides the thicker active layer, same stack architecture as for the solar cells characterized in section 4.1 was used. Identical devices are characterized with in-operando SKPM in the next section 4.4. Solar cell results on bilayer cells with 30 nm active layers are reported in section 5.1.2.

The solar cell samples were prepared at iL and transferred to Zeiss Oberkochen for FIB treatment under nitrogen atmosphere. Before the load to the Orion multi FIB microscope the samples had to be exposed to air. Also on the transfer to Heidelberg the samples were kept in ambient air. Characterization was performed as usually in the high vacuum of the Auriga setup. The solar cell samples treated with different FIB ions were identical and prepared within the same batch/deposition process. The SKPM measurements were performed using standard SKPM setup and parameter which were used also for measurements presented in the next section 4.4.

Bilayer solar cell stack as ion target

The free software package *SRIM* (Stopping and Range of Ions in Matter) was used to study the ion implantation in solar cell cross sections via FIB exposure [185]. Typical SRIM calculations are performed to determine ion penetration, phonon or damage profiles for certain materials or compounds. The ions impinge successively and with defined energy on the target surface and individual trajectories are simulated by means of a Monte Carlo algorithm. Layered stacks as targets can be defined in the graphical user interface (GUI), so that the assembly of the bilayer stack was straight forward. Organic layers were assumed to contain carbon only, ITO layers were modeled by a homogeneous intermixing of the three components In (74%), O (18%) and Sn (8%).

4 Electric potential distribution of F_4ZnPc/C_{60} small molecule organic solar ce

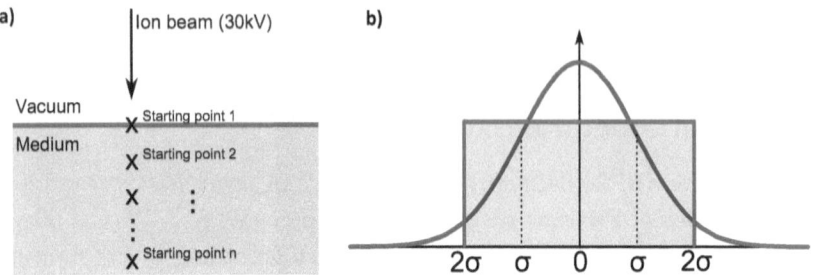

Figure 4.7: a) FIB milling is modeled with SRIM by a sequence of ions with starting points of increasing depth and constant initial kinetic energy of 30 kV. Via the number of ions released per starting point different sputter yields are modeled. b) Comparison between Gaussian (blue) and constant (gray) ion distribution over the 2σ range. Constant ion distribution in the limits of 2σ (gray area) is assumed here to calculate the density of implanted ions.

The milling approach with SRIM

The milling process was modeled by introducing FIB ions of identical incident energy with gradually increasing depth coordinate z. In figure 4.7 this approach is depicted: the ions are introduced in the stack with successively deeper starting points, where FIB ions cascade into the bulk material as in the case of FIB milling. Different sputter yields (i.e. the ratio of sputtered atoms per incident energetic ion, see section 3.1.4) are taken into account by the number of ions introduced per step on the z range: the higher the sputter yield, the less ions are necessary to achieve the same milling result. This functionality is not implemented in the SRIM GUI, but could be realized bypassing the GUI by means of tailored input data files. SRIM allows the use of input files simulating propagation of ions with arbitrary energies, starting coordinates and incident angles. Python scripts (shown in Appendix 7.1) were coded to generate appropriate input files taking into account layer thicknesses and different material sputter yields. All simulations were performed with ≈ 72500 incident ions of 30 kV energy. The different sputter yield of the FIB ions is taken into account by linear scaling of the extracted implantation profiles. Here only ion beams incidenting normally on the target are presented. The simulation of beams incidenting in 36° with respect to the target surface is straight forward to implement, but significantly more complex to evaluate.

4.3 The influence of FIB preparation on SKPM results

In test simulations we found that the impact of this simplification is minor considering the assumptions made for evaluation discussed in the following.

Extraction of ion implantation profiles

With the described input data SRIM simulations were run to simulate ion implantation in solar cell cross sections. Profiles reporting on the number and the densities of implanted FIB ions are extracted. The detailed calculation is documented in Appendix 7, here a short overview on the reasoning is given. Four steps were undertaken to estimate ion implantation from the simulation.

- From the FIB currents used in Ga FIB milling routines at the Auriga setup (see section 3.1.5 for details) we estimated the number of ions necessary to mill trenches as simulated with SRIM. For Ga FIB milling the simulated implantation profiles were calibrated with the empirical values. For Ne and He FIB milling the averaged difference in sputter yield Y was used to calibrate for the mismatch between simulation and actual milling process. Based on values presented in table 3.1 we used $Y_{He} \approx 10^{-2} \cdot Y_{Ga}$ and $Y_{Ne} \approx 0.2 \cdot Y_{Ga}$ in the calculations.

- Sputtering is not considered in the SRIM simulation. We assume that the ratio of FIB spot size and milling resolution is a meaningful measure to estimate ion implantation into the material matrix. Given typical values for He, Ne and Ga FIB spots, this corresponds to the implantation of every 4th/6th/4th FIB ion into the material matrix for He/Ne/Ga respectively.

- Our calculations are based on implantation profiles that give the azimuthally integrated number of ions implanted at a certain depth. A large fraction of these implanted ions is sputtered in the real FIB milling process, because not a spot but (typically) a rectangle is milled. The spot simulated here represents only one of identical spots that built up the last row in the actual FIB milling process. We assume that a quarter of the ions is indeed implanted in the FIB exposed cross section. With this we can give the total number of FIB ions implanted at a certain stack position.

	He$_{Ag}$	He$_C$	He$_{ITO}$	Ne$_{Ag}$	Ne$_C$	Ne$_{ITO}$
σ (nm)	50 ± 5	40 ± 4	60 ± 6	22 ± 2	10 ± 1	23 ± 2
δ (nm)	106 ± 49	200 ± 41	196 ± 88	30 ± 16	51 ± 16	52 ± 18

	Ga$_{Ag}$	Ga$_C$	Ga$_{ITO}$
σ (nm)	8 ± 1	4 ± 0.5	9 ± 1
δ (nm)	11 ± 6	23 ± 6	19 ± 11

Table 4.2: Radial variance σ of 30 kV FIB-target material combinations. σ refers to the radial spreading of the ions in the material. The penetration depth δ gives both maximum and peak width of the Bragg peaks in the material. The values refer to FIB ions normally incidenting on the target surface.

- The distribution of the ions implanted in the matrix varies strongly for the different FIB ions. The (square root of the) radial variance determines the radial spreading of ions from the FIB beam into the target material matrix. The values obtained in SRIM simulations are presented in table 4.2. We find that He ions penetrate 40 nm radially into the organic matrix, Ga (Ne) ions only 4 nm (10 nm). To account for this in the estimation of the implantation density, we assume that the density of implanted ions is constant in the range of 2σ around the beam and zero out of 2σ. Figure 4.7 b) depicts the approach. With this we can estimate the density of implanted ions versus the z position in the bilayer solar cell stack.

4.3.3 Results of SRIM simulations: ion implantation profiles

In this section we present simulation data on the ion implantation into the organic matrix of solar cell active layers during cross section exposure by means of FIB milling. After a short discussion on the impact of the input file design on the simulation output we present a comparison of milling induced ion implantation for FIB milling with He, Ne and Ga ions. It is commented on the ion-target interaction both within the different layers of the solar cell

4.3 The influence of FIB preparation on SKPM results

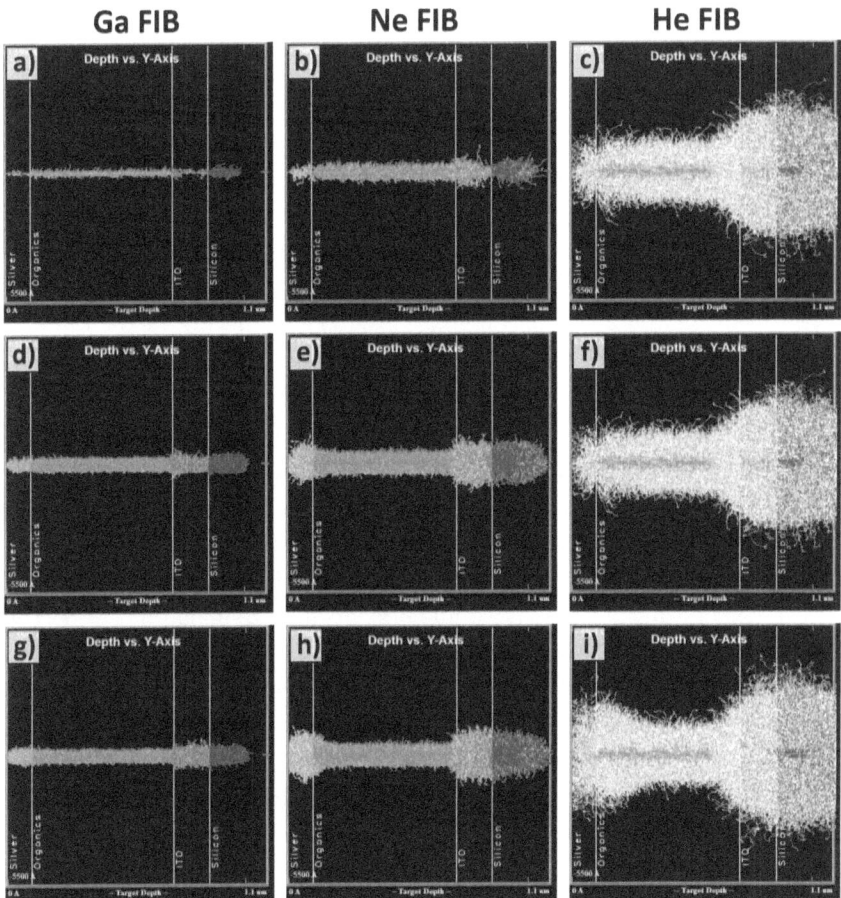

Figure 4.8: Top view on ion trajectories from FIB milling SRIM simulations. The solar cell materials are transparent here, so that only the ions and their trajectories are visible. In the left/middle/right column: Ga/Ne/He. a)-c) SRIM simulations including different material sputter yields. d)-f) simulations not including different material sputter yields. g)-i) Illustration of the different ion milling efficiency: the presented Ga (g) and Ne (h) simulations lead to the same milling result as the He FIB spot (i).

stack and between the different ions in use. All simulations were performed with ions of 30 kV incident energy.

Ion trajectory profiles: the qualitative picture

In figure 4.8 we present top views on simulated trajectories of individual Ga, Ne and He ions injected along the solar cell stack. The spot milling in a bilayer organic solar cell consisting of 100 nm Ag, 600 nm organics and 150 nm ITO on a glass (here: Si) substrate is simulated.

The first row shows the results of a simulation using ≈ 72500 Ga, Ne and He FIB ions each and taking into account the different sputter yield values for the respective ion-target combinations. The properties of the different ion-target material combinations lead to varying ion implantation. The strong radial spreading of the lighter FIB ions Ne and He in comparison to Ga as well as the correlation between higher sputter yield and higher transverse scattering in the metal-like layers are clearly visible. Also an inhomogeneous Ne and He distribution along the depth coordinate is visible, especially in the Ag top contact. In the following data from these simulations are used to estimate the extent of ion implantation in the solar cell cross section.

In the second row we present results of spot milling simulations neglecting the material dependency of the ion sputter yields. In the trajectory plots shown here the results look similar to the results presented above. Neglecting the higher sputter yields for Ag and ITO, we find a higher number of ions deposited there. However, when quantifying the results in the next section we find significant deviations in the ion distribution with respect to the previous case.

The comparison in the third row focuses the performance of the different focused ion beams. We present ion trajectory plots simulating Ga, Ne and He spot milling. Relating to the FIB spot obtained with He ions in the first row, we simulated the distribution of Ga and Ne ions which would lead to the same milling performance as achieved by He ions. For this, the number of incident Ga and Ne ions was related via the different sputter yields to the number of He ions. In the He milling simulation 72500 incident ions were injected throughout the layer stack, whereas 3325 Ne (730 Ga) ions were used to obtain the same effective milling with Ne (Ga). The results show the huge effect of the differing sputter yields: whereas there is only a slight ion

4.3 The influence of FIB preparation on SKPM results

contamination in the Ga milled sample, especially the He milled sample is permeated with FIB ions to a very high extent.

Ion implantation profiles: the quantitative picture

We estimated the implantation of FIB ions in the solar cell cross sections from SRIM data as outlined in section 4.3.2. The results based on data from the simulations presented in figure 4.8 are compared in figure 4.9.

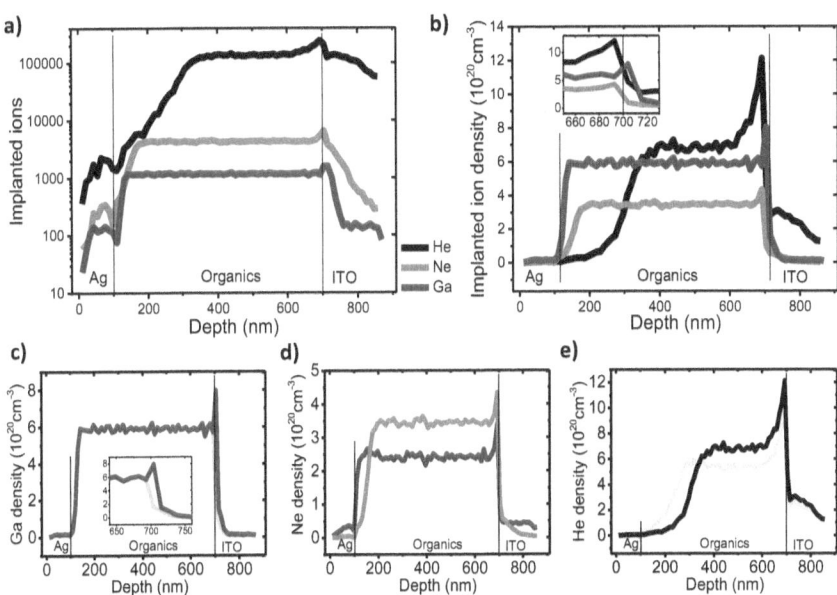

Figure 4.9: Depth profiles of He (black), Ne (red) and Ga (blue) ions implanted in the solar cell cross section calculated from SRIM simulations. Insets highlight the interface between organics and ITO bottom contact. The different layers of the stack are marked by vertical lines. a) Total number of implanted ions. b) Density of implanted ions. Second row: comparison between simulations including/not including the different material sputter yields. Including profiles are colored as above, profiles with neglected sputter yield are given in c) light blue (Ga), d) violet (Ne) and e) gray (He).

In figure a) the depth profiles of implanted ions are compared on a logarithmic scale. The picture is dominated by large differences concerning the amount of ions necessary for achieving the same FIB milling effect. The relations $Y_{He} \approx 10^{-2} \cdot Y_{Ga}$ and $Y_{Ne} \approx 5 \cdot Y_{Ga}$ lead to ion implantations in the organics of He (Ne) that are about 100 (5) times higher than for Ga. Besides that we find large inhomogeneities both along the bilayer stack and in the active layer. The differences in implantation between the layers is caused by the about 10-fold higher sputter yield of ions in Ag and ITO than in the organics: because more ions are necessary in the milling process of the latter, the number of implanted ions is higher. The inhomogeneities within the active layers stem from the penetration depths δ given in table 4.2: the penetration depths of He (Ne) ions are in the range of the layer thicknesses in the stack and with 200 ± 41 nm (50 ± 16 nm) very high for ions in the organics. This leads to a valley after the Ag top contact expanding into the organic layer by about δ nm. This behavior becomes less pronounced with higher ion mass.

In figure b) the density of ion implantation in the stack is presented in linear scale. The large differences in the number of implanted ions found in a) is counterbalanced here by the stronger radial variance σ of the Ne and especially the He ions. As described in 4.3.2, for the estimation of ion density we assume a rectangular radial ion density along the beam with a width of 2σ. The radial spreading of the FIB ions is documented in table 4.2: we find approximately $\sigma_{He} \approx 10 \cdot \sigma_{Ga}$ and $\sigma_{Ne} \approx 3 \cdot \sigma_{Ga}$ for all materials. This means that there is strong ion implantation within the first 80/20/8 nm of cross sections prepared by He/Ne/Ga FIBs. There are large inhomogeneities concerning the implantation density within the organic layer. In table 4.3 the ion implantation density is given for spots on the top, the center and the bottom of the organics in the device. For He FIB milling, we find inhomogeneities from top to center (bottom) of a factor 140 (240) within the organics. For Ne FIB milling, the density at these positions differs by a factor of 12 (14). For Ga FIB milling, the density is inhomogeneous by a factor of 15 only at the very top of the organic layer. The density towards the bottom contact does not vary significantly from the bulk. As visible in the inset of figure 4.9 b) the ion density peaks here, in contrast to the peaks of the He and Ne profiles, in the ITO bottom electrode rather than in the organic active layer.

In the second row of figure 4.9 we present a comparison of ion density profiles extracted from the simulations shown in the first row (considering sputter yield characteristics of the stack materials) and the second row (neglecting

4.3 The influence of FIB preparation on SKPM results

Stack position	He (10^{20}cm^{-3})	Ne (10^{20}cm^{-3})	Ga (10^{20}cm^{-3})
$x = 110$ nm	0.05	0.3	0.4
$x = 500$ nm	7	3.5	6
$x = 695$ nm	12	4.3	6

Table 4.3: Ion implantation density of FIB ions at particular depth coordinates within the stack. The first coordinate refers to the very top, the second to the center and the third to the bottom of the organic layer. Data is extracted from the graphs in figure 4.9 b).

sputter yield characteristics) of figure 4.8. The evaluation procedure of the profiles was identical. To account for differences between the sputter ions, the profiles were weighted with the average difference in sputter yield between Ga and Ne (factor 5) and He (factor 100). Although the simulations (for the same FIB ions, i.e. column-wise in figure 4.8) appear very similar in the trajectorial picture, here differences emerge. Of course the main differences appear between the implantation in the organic and the electrode layers (caused by their large difference in sputter yield), which is only of minor interest here. However, we find both varying density profile shapes and slight shifts in values also within the active layers. In the Ga profile the peak in density at the top of the ITO layer does not appear when sputter yield is neglected (see the inset). The regions of lowered implantation at the top of the active layer is underestimated both for Ne and He milling when sputter yield between the materials is neglected. Furthermore this neglect leads to a slight underestimation of the implantation density on the plateau within the active layer of about 30% (25%) in the Ne (He) profiles.

Conclusion: SRIM simulations

By means of simulating ion implantation in the solar cell stack we found that FIB milling leads to an inhomogeneous ion density distribution both along the entire solar cell stack and within the organic active layer. These effects are more pronounced for the lighter FIB ions He and Ne than for Ga. The ion densities right at the top and the bottom of the active layer vary by a factor of 15/14/240 for Ga/Ne/He ions. So also using the standard FIB ion

Ga for milling leads to significant inhomogeneities within the active layer that could influence results in cross-sectional studies with high lateral resolution. Although implantation densities on the same order for all ions were found, we assume the doping impact to be higher for the electrically more active Ga ions. In experiment it is possible that a significant amount of the lighter Ne and especially He ions leave the sample cross section due to outgasing into the high vacuum of the microscope chamber. This is probable at least for the ions implanted at the very surface of the cross section. However, we have no measure to quantify and account for this here.

In a comparison of ion implantation profiles extracted from simulations accounting/not accounting for different sputter yields of the materials in the stack, we could demonstrate that neglecting the variation of sputter yields leads to an underestimation of ion inhomogeneities. Nevertheless, qualitatively these simulations lead to same findings of strongly inhomogeneous FIB ion implantation within the organic active layer.

4.3.4 Results of SKPM studies: electric potential profiles

In this section data from SKPM studies on organic solar cell cross sections exposed using FIBs with different sputter ions are compared. For details on SKPM and the interpretation of its results it is referred to section 3.1.2. As described in the experimental details 4.3.2, FIB milling with He and Ne ions was performed at Zeiss in Oberkochen. Ga FIB milling as well as SKPM characterization was performed at iL in Heidelberg.

Further details on sample preparation, performance of measurements and the BRR AFM SEM experimental setup are found in [119, 173, 175].

In figure 4.10 SKPM depth profiles for the differently FIB prepared cross sections of identical solar cell samples along with comparisons of implantation and SKPM profiles are presented. The solar cells were characterized under short circuit (SC) conditions (0 V bias voltage applied).

We find that all potential profiles exhibit at least in parts the bowl behavior discussed in the state of the art 4.3.1. Both preparation with Ne and with Ga FIB leads to the observed shape corresponding to work functions in the stack that are highest for the organic materials in the active layer. Based on KP results from a step-by-step characterization of bilayer stacks with F_4ZnPc

4.3 The influence of FIB preparation on SKPM results

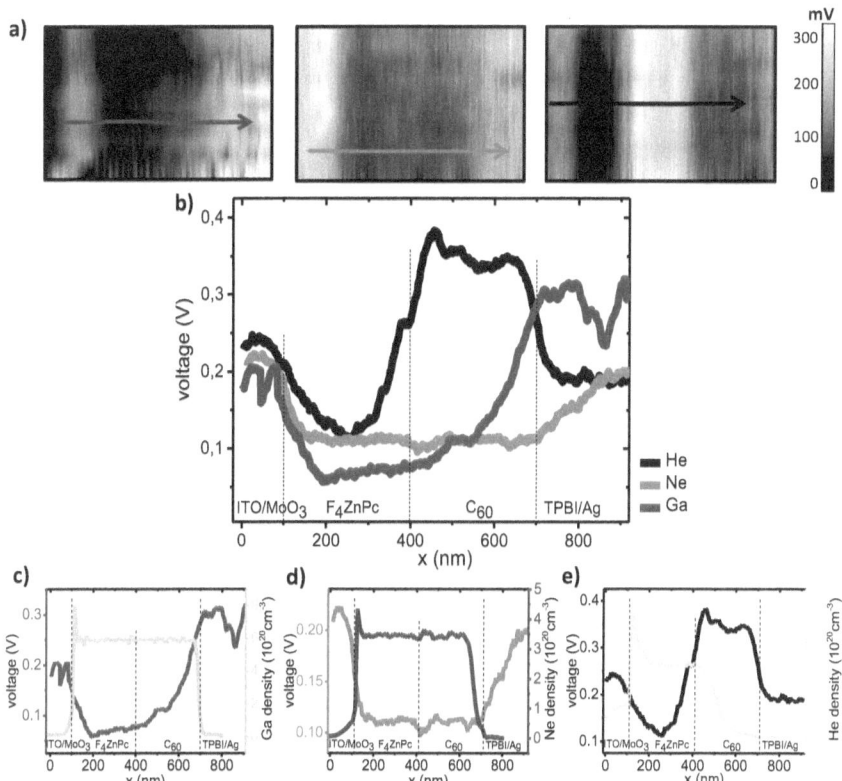

Figure 4.10: SKPM results of solar cell samples under SC conditions (0 V applied) prepared with different FIB ions. Arrows indicate where profiles were extracted. a) SKPM potential maps of Ga/Ne/He FIB milled samples (1.4 µm×16 nm each). b) SKPM profiles of organic solar cells prepared by Ga (blue), Ne (red) and He (black) FIB ions. The vertical lines indicate the interfaces in the device. c)-e) Ion implantation and SKPM profiles are correlated for (c) Ga (light blue/blue), (d) Ne (violet/red) and (e) He (gray/black) ions. Left y-axis: CPD from SKPM measurements; right y-axis: ion implantation density. The implantation profiles are identical to the ones in figure 4.9 b), but presented here in opposite x-direction for comparability.

and C_{60} presented in section 4.1, we expected a work function (wf) evolution across the stack with a high wf of the hole extracting contact (ITO/MoO$_3$), followed by a lower wf of the F$_4$ZnPc layer and a slightly higher wf for the C_{60}. Towards the electron extracting contact a further lowering in wf was expected. This is not observed in the measurements here.

Ne FIB prepared samples The profile of the Ne FIB prepared solar cell does not show significant deviations from the Ga FIB prepared cell. There is a slightly lower work function detected for the Ag top contact. However, such variations are not unusual in SKPM measurements of this type and can appear also on identically prepared samples [5]. The trend in the active layer follows the observations on Ga FIB treated samples discussed in 4.3.1, indications towards a significant influence of the inhomogeneous ion distribution in the active layer are not found.

He FIB prepared samples Only the profile of the He FIB prepared sample exhibits varying wf in the active layer region, but the trend is opposite than expected: whereas in the step-by-step KP experiment we found a lower wf for the F$_4$ZnPc layer ($\phi_{F_4ZnPc} = 4.7$ eV, $\phi_{C_{60}} = 4.9$ eV), here we find the work function about 200 meV lower in the C_{60} layer. The increase in wf for the F$_4$ZnPc layer could be caused by p-doping of the material. When comparing the SKPM and the implantation profiles in figure 4.10 e), we find that the rise in the SKPM profile expanding from 300 to 450 nm coincides with the decrease in the density of implanted ions. This raises the question whether the increased wf (lower SKPM signal) in the F$_4$ZnPc layer of the He prepared cross section could stem from the enhanced He implantation. As discussed in the state of the art 4.3.1, in graphene samples treated with He FIB n-doping was found [182]. We observe p-doping in the F$_4$ZnPc layer, because opposite than expected ϕ_{F_4ZnPc} is found higher than both ϕ_{ITO/MoO_3} and $\phi_{C_{60}}$. In this layer a significantly higher He implantation is expected from simulations. However, the peak in the He implantation density at the bottom of the active layer ($x \gtrsim 100$ nm) did not lead to an additional increase of ϕ_{F_4ZnPc}, but this could be caused due to the limited lateral resolution in SKPM also.

---

[5] The focus was on fast characterization of all samples at the same time, so it was renounced on time-consuming optimization steps.

Conclusion: SKPM studies

We discussed SKPM results gathered on organic solar cell cross sections exposed by different FIB ions in the light of ion implantation profiles extracted from SRIM simulations. We found only minor deviations in the profiles of Ga and Ne FIB prepared samples which were limited on the electrode region. In the active layer region the typical bowl behavior was found. Relative to $\phi_{\text{ITO/MoO}_3}$ and $\phi_{\text{C}_{60}}$, the He FIB prepared sample exhibited a significantly increased F_4ZnPc work function which could be induced by the stronger p-doping of this layer because of inhomogeneous He ion implantation or beam damage found in SRIM simulations.

In conclusion, the goal to obtain high quality 0 V-SKPM profiles by performing Ne or He FIB instead of Ga FIB preparation was not achieved. Be it due to the exposure to ambient air after the cross section preparation or due to contamination by He or Ne implantation, we did not obtain profiles exhibiting the expected trends derived from the KP studies in 4.1 or literature.

4.4 In-operando SKPM studies on OPV cells with varied hole extracting contacts

In this section in-operando cross-sectional SKPM studies on F_4ZnPc/C_{60} bilayer solar cells are presented. Devices with (decent) ITO and (good) ITO/MoO$_3$ hole extracting contacts were characterized to study the impact of controlled contact manipulation on the electric potential distribution at the nanoscale. The results are discussed in the light of findings from IV characterization (see 5.1.2) and complementary analytical studies with Kelvin Probe (see 4.1) and photoelectron spectroscopy (see 4.2). The solar cell cross sections were Ga FIB prepared as discussed in the previous section 4.3 and characterized with SKPM under working conditions, i.e. under illumination and applied bias voltage. Details on SKPM and in-operando studies based on this method are discussed in section 3.1.2.

4.4.1 State of the art

SKPM was introduced in 1991 from Nonnenmacher et al. [79] and became by now a standard characterization tool especially in the field of material science. First cross-sectional SKPM studies on GaAs multiple quantum well structures were published by Chavez-Pirson [98], cross-sectional in-operando studies on light emitting diodes by Shikler et al. followed 1999 [186]. Since then, most relevant types of solar cells were characterized by cross-sectional SKPM [187–194]. Already in the first works in this field it became evident that the electric potential distribution determined with SKPM at the surface of the (here) cleaved devices does not necessarily match the actual bulk values. In these crystalline samples modifications emerged caused by band bending effects at the exposed cross sections. Although surface band bending effects are less relevant in the organic solar cells characterized here, contamination induced by exposure to ambient air or the FIB preparation can lead to deviations from bulk values (see the discussion in 4.3.1). Because of this, for large parts of the discussion on our data we rely on an evaluation procedure proposed by Bürgi et al. for in-operando SKPM studies in 2001: relative potential profiles are considered rather than the as-measured profiles [195]. They are obtained subtracting the 0 V-profiles (device short circuited) from the profiles gathered under applied bias voltages. Besides the benefit of eliminating surface induced artifacts this has also the advantage that the net effect of the bias voltage applied is obtained, which is often the most interesting parameter in theses studies (see the discussion in 3.1.2.1). This procedure is widely accepted in the SKPM community and applied in several publications on both inorganic [171, 187] and organic solar cells [12, 13]. By now, some in-operando cross-sectional SKPM studies on organic solar cells were published. Here a brief overview on these publications is given.

Lee and Kong et al. were 2011 the first who published cross-sectional SKPM studies on organic solar cells. They investigated cleaved P3HT:PCBM BHJ solar cells under short circuit (SC) conditions and found the built-in electric field in the device confined to the contact regions with the bulk of the BHJ nearly field-free [193]. However, lacking an effective electron extracting contact (Al top contact was applied directly on the BHJ, so no blocking/extraction layers were used) the characterized devices did not meet

4.4 In-operando SKPM studies on OPV cells with varied hole extracting contacts

the standards of state-of-the-art organic solar cells.

In 2012 they presented SKPM studies with very high lateral resolution (active layers were < 150 nm) on cleaved state-of-the-art BHJ solar cells in conventional architecture (using TiOx/Al electron extracting contacts) under applied bias voltages [194]. They found that a large share of the potential drops in the contact region, namely at the electron extracting top contact. Here, electric field strengths of $4 \cdot 10^5$ V / cm were found, with only $3 \cdot 10^4$ V / cm in the BHJ bulk (at -0.8 V bias voltage applied). For a bias voltage of $+0.4$ V corresponding (approximately) to the maximum power point (MPP) of the cell, they found the bulk of the device almost field-free ($< 1 \cdot 10^4$ V / cm). They stated that this field-absence would ensure the good performance of the device, because diffusion driven photocarriers could easily reach the respective contacts. However, the validity of the results is questionable because the influence of illumination was not addressed.

Saive et al. published in 2013 an extensive cross-sectional SKPM study on both conventional and inverted P3HT:PCBM organic solar cells characterized under illumination and applied bias voltages [12]. For cross section exposure Ga FIB was used. Besides the bowl-shaped profiles for devices under SC (broadly discussed in the previous section 4.3.1) they found very different behavior under bias voltage for devices in conventional and inverted architecture: whereas in the conventional devices the applied potential dropped almost at the contact regions only (to approximately the same extent at both contacts), in the inverted devices the potential drops linearly all over the BHJ. This was explained by the superior contact properties in the inverted devices. However, the explanation was not cross-checked with IV characteristics of the inverted devices. The conventional BHJ solar cells were also characterized under illumination: by applying different contact scenarios it was demonstrated that the photovoltage under open circuit (OC) conditions drops at the electron extracting contact only.

Chen et al. published in 2015 a similar study using argon FIB to expose the solar cell cross sections [196]. Results on P3HT:PCBM and P3HT:ICBA BHJ solar cells in conventional architecture under bias voltages and illumination were compared. Although partly contradicting our results from cross section preparations with Ne and He FIB presented in the previous section 4.3, according to Chen et al. the Ar FIB cross section exposure allows for the extraction of SKPM profiles of organic solar cells under SC that match the expectations from theory. As reported by Saive et al., also here the large part of the photovoltage drops at the electron extracting contact when the device is operated at OC conditions. Under applied bias voltage, they observe an approximately constant electric field all over the BHJ. A helpful *bias compensation method* is proposed for the estimation of the built-in potential in the device: because convolution effects prevent the extraction of meaningful values of the built-in potential from SC profiles, a bias voltage V_{app} is applied to bring anode and cathode to the same CPD in the SKPM measurement, so that yields $V_{bi}=V_{app}$. However, besides the fact that many achievements claimed in this report are common sense in the SKPM community, many of the measurements presented suffer from limited lateral resolution in the order of the active layers. The deconvolution approach is presented in a misleading way, pretending that a full reconstruction of the actual (real) potential distribution from the convoluted SKPM data is possible, which is not the case (see the discussion on convolution in SKPM in section 3.1.2).

Saive et al. presented a cross-sectional SKPM study on P3HT:PCBM BHJ solar cells with regular diode and s-shape IV characteristics [13]. By comparing SKPM measurements on regular and s-shape solar cells performed under applied bias voltages they could demonstrate that the s-shape behavior is caused (at least in this case) by the poor quality of the electron top contact. Whereas the applied potential drops all over the BHJ bulk in the case of the regular diode, the entire potential drops at the top contact in the case of the s-shaped solar cell.

Scope of this work

In this work we focused on two issues we found not addressed in literature so far. Besides the mentioned paper on s-shape solar cells in all publications BHJ solar cells were studied. These are characterized by DA blends intermixed on the nanoscale, so that processes of charge generation right at the DA interface could not be accessed because of the limited lateral resolution in SKPM. Here we characterize organic bilayer solar cells with film thicknesses large

4.4 In-operando SKPM studies on OPV cells with varied hole extracting contacts

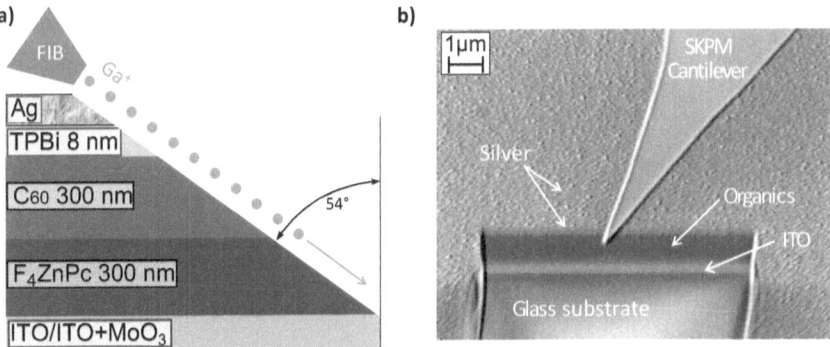

Figure 4.11: a) Sketch of bilayer solar cell stack and FIB preparation for cross-sectional SKPM characterization. Plasma treated ITO and MoO$_3$ coated ITO are used as bottom contacts, the rest of the stack is identical. b) SKPM measurement in action: the conductive cantilever is approached to the FIB exposed solar cell cross section and scans it in vertical direction.

enough to allow detailed insights in charge generation processes at the DA interface. Another goal of this work was to study the impact of a controlled contact variation on the nanoscale potential distribution in the device. The studies published so far on BHJ solar cells could not address this point. The complex morphology of the BHJ can vary from homogeneous intermixing to distinct enrichment of one of the components towards the bottom or top contact depending on annealing procedures and substrates in use. Thus in the above-mentioned SKPM studies the exact BHJ morphology and with it the actual interfaces present at the contacts were unclear. This is overcome here by a vacuum processed bilayer system whose contact configuration is defined and well-understood due to interface studies with the complementary methods KP (see 4.1) and PES (see 4.2).

4.4.2 Experimental details

The solar cell stacks characterized here are depicted in figure 4.11. As well shown is the FIB exposure of solar cell cross sections and a SEM image of the SKPM measurement configuration. The two hole extracting contacts (plasma treated ITO and ITO/MoO$_3$) already discussed in the XPS/UPS study presented in section 4.2 were realized in these solar cells.

Both substrates were cleaned in an ultrasonic bath of Acetone and Isopropanol for 10 min each, followed by a O_2 plasma treatment of 5 min. The samples were transferred to the UHV within 15 min after plasma treatment. Here F_4ZnPc/C_{60} bilayer solar cells with thick active layers of 300 nm each and an electron extracting TPBi/Ag top contact were prepared according to the standard procedures described in section 3.2. Besides the varied hole extracting bottom contact the solar cells were identical.

The contacting of the solar cells and the transfer to the high vacuum (HV) of the BRR SEM AFM had to be performed in ambient air. After transfer to the HV chamber of the microscope the cross sections were exposed with Ga FIB. In measurement configuration the operation of the solar cells was controlled by IV characterization. In automatized routines bias voltages were applied and the cross section scanned. By locking the current response of the cell, proper operation of the samples could be ensured. The solar cell cross section and the cantilever were characterized after a successful SKPM measurement series by means of SEM to cross-check layer thicknesses found in SKPM and to control that the cantilever was free of damage.

For details on the setup see section 3.1.5 or the theses of Rebecca Saive [119] and Tobias Jenne [175].

We note that the results discussed here were consistently found for all samples investigated and the findings are based on at least two identical measurement series performed at different times on identical solar cell samples of different batches.

4.4.3 Results: In-operando SKPM studies

In this section results of SKPM studies on F_4ZnPc/C_{60} bilayer solar cells under working conditions are presented. First, bilayer solar cells are compared under illumination to study features induced by the charge generation at the DA interface and improved hole extraction at the MoO_3 coated ITO contact. SKPM studies under applied bias voltages are discussed to study the internal field evolution of the devices and to draw conclusions on conduction mechanisms and mobilities in the F_4ZnPc and C_{60} layers. Results on the performance of F_4ZnPc/C_{60} bilayer solar cells with identical contact architecture as used here but with optimized layer thicknesses are presented in section 5.1.2: samples with MoO_3 coated ITO (plasma treated ITO) contact

4.4 In-operando SKPM studies on OPV cells with varied hole extracting contacts

exhibited PCEs of $\approx 1.6\%$ ($\approx 1.2\%$) with $V_{oc} \approx 700\,\text{meV}$ ($V_{oc} \approx 640\,\text{meV}$), $j_{sc} \approx 3.8\,\frac{\text{mA}}{\text{cm}^2}$ ($j_{sc} \approx 3.4\,\frac{\text{mA}}{\text{cm}^2}$) and comparable FF of $\approx 60\%$.

To account for convolution induced reductions in the SKPM voltage contrast, extracted data is scaled to the actual bias voltages applied. For example, if we obtained a voltage contrast of 0.8 V between anode and cathode in the (SC corrected, relative) SKPM profiles at a device under 1 V bias voltage, we multiplied the entire profile by a factor of 1.25 to end up with a contrast between anode and cathode of 1 V. This evaluation procedure was proposed by Bürgi et al. [195] in 2002, in the meantime it is well-established in the SKPM community [91, 197, 198]. It ensures comparability between different measurement series (with different convolution factors) and meaningful estimations of field strengths in the device. In our experiments we found that the contribution of cabling and contact resistances outside the devices is negligible. In the given cantilever-sample geometry the cantilever faces the grounded Ag top contact at constant height during the whole measurement, so that the effect of convolution can be assumed as constant here and the application of the discussed evaluation is justified.

4.4.3.1 Results 1: Studies on illuminated devices

In figure 4.12 the main results of a measurement series on bilayer solar cells with MoO_3 coated ITO and plasma treated ITO are presented. By SEM observations after the measurements the sample as well as the cantilever were found undamaged. In high-resolution SEM the layer thicknesses were controlled. In the cell with plasma treated ITO contact a certain mismatch to the intended geometry was found: the layer thickness of F_4ZnPc was $330 \pm 30\,\text{nm}$, for C_{60} $260 \pm 30\,\text{nm}$ was found. For the MoO_3 coated ITO cell layer thicknesses of $300 \pm 30\,\text{nm}$ were found for both F_4ZnPc and C_{60}. We have no indications that the mismatch found limits the scope of the findings made here.

With IV curves after FIB preparation the operation of the solar cells was checked and the open circuit voltage determined: we found $V_{oc} = 560\,\text{mV}$ ($V_{oc} = 530\,\text{mV}$) for the MoO_3 (plasma ITO) cell and a working point of $V_{mpp} \approx 350\,\text{mV}$ ($V_{mpp} \approx 400\,\text{mV}$) respectively. However, the white LED light source used illuminates the solar cells in a non-standardized way and although the LED was driven under identical conditions, slight variations in

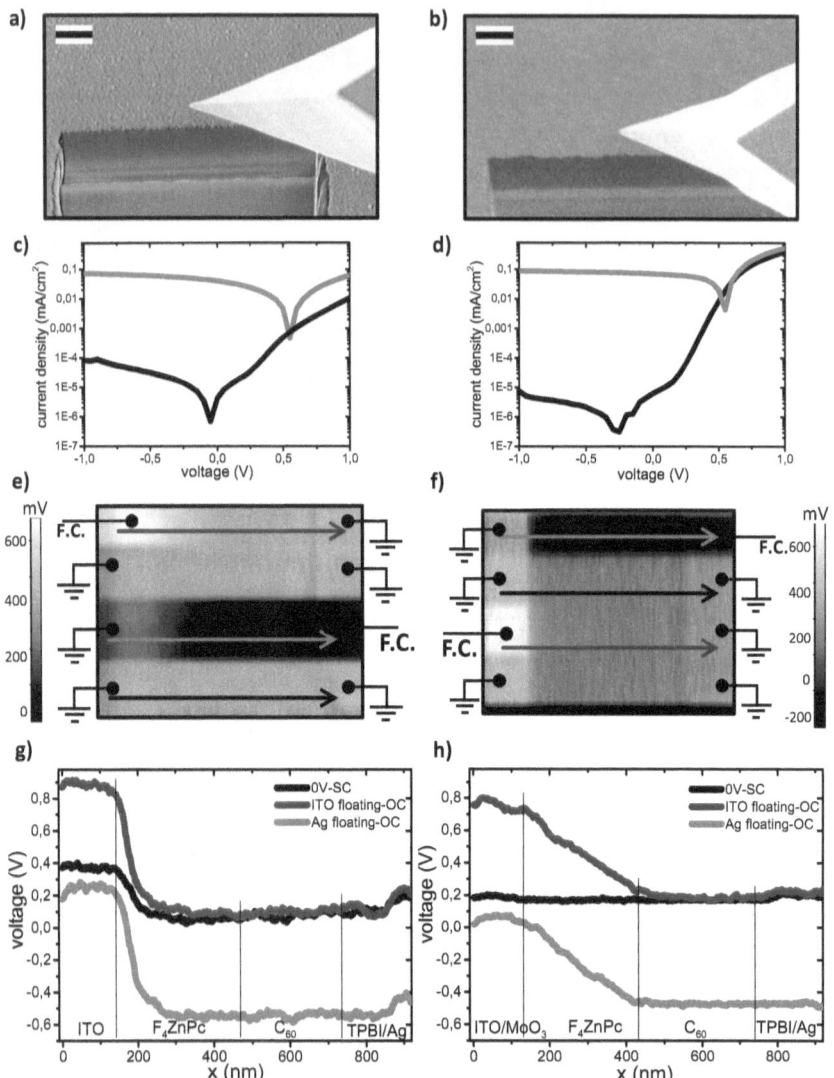

Figure 4.12: SKPM measurement series on bilayer solar cells with ITO/MoO$_3$ (left column) and plasma treated ITO (right column) hole extracting contact. a), b) SEM micrographs of FIB milled cross section and non-damaged cantilever after the measurement series. Scale bar corresponds to 1μm. c), d) IV curves of the samples after FIB preparation. e), f) CPD micrographs (1.2 μm × 80 nm each) measured with SKPM under illumination and different contact scenarios: SC conditions (both contacts grounded); OC conditions (disconnected, i.e. "floating" contact = F.C.). g), h) SKPM profiles extracted from the SKPM potential maps in e), f) at positions marked by the arrows. Vertical lines indicate the interfaces in the devices.

4.4 In-operando SKPM studies on OPV cells with varied hole extracting contacts

illumination intensities can not be excluded.

The SKPM maps for measurements under illumination and different contact scenarios of both samples are presented. During the scan of the image we changed the contacts to create SC and OC conditions. In SC 0 V is applied, i.e. both contacts are grounded (here sketched by the earth ground symbol). In OC the circuit is opened by disconnection of one of the contacts. This is sketched here by the "floating contact" (F.C.). We find that the disconnection of the electron extracting contact under illumination (red arrow) induces a negative shift of the disconnected contact in the CPD signal, corresponding to a lowering in the distance between Fermi and vacuum level (i.e. electron enrichment at the contact). When disconnecting the hole extracting contact, we find a positive shift of this contact in the CPD signal (i.e. hole enrichment). As expected, electrons/holes accumulate at the respective disconnected contact.

For quantitative evaluation of the data single line profiles are extracted from the CPD maps. The profiles presented here are scaled to the open circuit voltages V_{oc} extracted from IV curves.

The convolution effect in the measurements can be observed here very clearly: when the silver top contact is floating (red profile), its potential decreases. As can be seen on the slight negative shift of the ITO contact ($\approx 100\,\mathrm{mV}$) this impacts also the hole extracting contact, although the latter is constantly on same potential (grounded). This demonstrates the interaction of the cantilever with the entire Ag top contact.

Impact on the DA interface

Opposing to our expectations we did not find any specific features indicating charge generation at the DA interface built up by F_4ZnPc and C_{60}. We think that this is a method-inherent issue and not connected to the potentially poor performance of the devices, because both in IV and SKPM measurements the functionality of the solar cells was demonstrated (see the following discussion). The same was found by Rebecca Saive on solution processed bilayers [119]. However, in her study it could not be assured that there is still a certain intermixing of the P3HT and PCBM layers. Here this option is excluded.

Comparison of contact qualities

In figure 4.13 relative profiles of the SKPM measurements from figure 4.12 are presented. The relative profiles are obtained by subtracting the 0 V-profiles (SC) from the profiles under OC conditions. For better comparability, the Ag floating-profile was shifted on the voltage (y) axis. This shift has no physical meaning in the picture of the relative profiles here. Also shown are profiles where $V = V_{oc}$ was applied (SKPM maps not shown here).

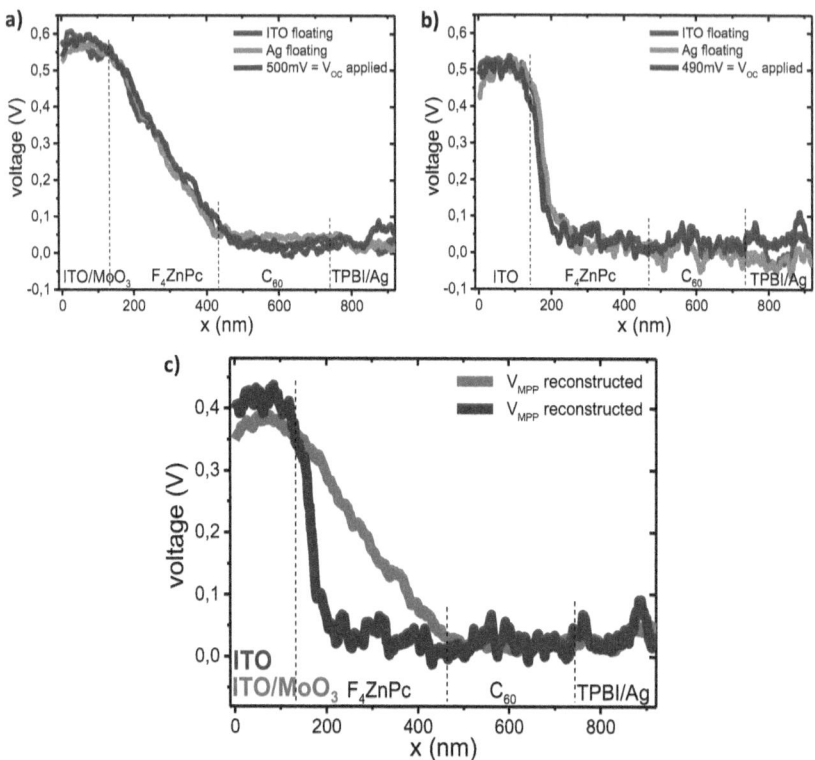

Figure 4.13: Relative SKPM profiles extracted from the as-measured profiles of the MoO_3 (a) and the plasma treated ITO (b) sample by subtracting the 0 V-profile. OC conditions induced by floating ITO (Ag) contact are given in blue (red). The Ag floating-profiles were shifted on the voltage axis. Also a profile with a bias voltage of $V = V_{oc}$ is shown (violet). c) The potential distribution of the cells at their working point $V = V_{mpp}$ reconstructed from a) and b). MoO_3 (ITO) sample given in green (brown).

4.4 In-operando SKPM studies on OPV cells with varied hole extracting contacts

Potential distribution under OC conditions We find for both devices that the relative potential profiles of both the OC case induced by floating hole (ITO) and electron (Ag) extracting contacts overlap. In line with the reports of Saive et al. [12] and Chen et al. [196] this applies also to profiles obtained from SKPM measurements with $V = V_{oc}$ applied.

For the MoO_3 coated ITO contact we find a linear (photo-) voltage drop over the entire F_4ZnPc donor layer and the C_{60} acceptor field-free. We attribute this behavior to the good hole extraction properties of the MoO_3/F_4ZnPc contact interface already found in UPS studies (see section 4.2) and the poor hole mobility of F_4ZnPc compared to the electron mobility in C_{60}. As discussed in the material section 3.2.1 typically hole mobilities in Phtalocyanines are significantly lower than electron mobilities in C_{60}. Here, the good hole and electron extraction of the contacts preventing the accumulation of charge carriers at the contact interfaces emphasize the bulk properties of the active materials. Under OC conditions, the diffusion driven photocurrents originating at the DA interface have to be compensated by drift currents from the contacts of same magnitude and opposite direction. In equilibrium these currents have same magnitude both in the donor and the acceptor layer. To reach this, the field strength has to be significantly higher in the poorly conducting F_4ZnPc layer. In conclusion, we find a distinct bulk limitation of the bilayer cells with MoO_3/F_4ZnPc hole extracting contact.

For the plasma treated ITO contact we find that the entire photovoltage built up in the solar cell drops in direct vicinity of the hole extracting contact only. From XPS studies of the plasma treated ITO/F_4ZnPc contact we know that a significant share of the first F_4ZnPc monolayer decomposes because of the high ITO chemical reactivity (see section 4.2). It is reasonable to assume that this goes along with a limited hole extraction, manifested for example in a finite recombination velocity (discussed for BHJ solar cells in [166]) . Therefore we attribute the observed behavior to (donor-sided) hole accumulation which screens the photofield in the solar cell. In contrast to the good hole contact with linear decreasing potential all over the donor layer we find that here the solar cell exhibits distinct contact limitation.

Potential distribution at maximum power point V_{mpp} Having measured the potential profiles of the illuminated cells under SC and OC conditions by applying $V = V_{oc}$, we can reconstruct the potential distribution in the cells at their working point, i.e. at an applied voltage of $V = V_{mpp}$: for this, we estimate V_{mpp} from the IV curves presented in figure 4.12 (MoO_3 sample: $V_{mpp} \approx 350\,mV$; ITO sample: $V_{mpp} \approx 400\,mV$) and reconstruct the relative V_{mpp}-profiles with the relation:

$$V_{mpp} - \text{relative profile} = \frac{V_{mpp}}{V_{oc}} \times (V = V_{oc}) - \text{relative profile}$$

from the relative profiles with $V = V_{oc}$ applied. Assuming continuity in the evolution of the potential, this is justified here: On the one hand, 0 V-profiles (SC) correspond to horizontal profiles in the relative profile-picture in figure 4.13. On the other hand, we know the potential distribution for an applied voltage of $V = V_{oc}$. Because of this, the V_{mpp}- relative profiles can only (i) be located between the two profiles mentioned and (ii) have to be assembled of both the shape of the $V = V_{oc}$- and the 0 V-relative profile (horizontal here) for reasons of continuity. Therefore the V_{mpp}-relative profile can be obtained by simply weighting the $V = V_{oc}$-relative profile with the ratio of V_{mpp} and V_{oc}. The profiles obtained are presented in figure 4.13. Of course, the trends discussed above also dominate the picture here.

Conclusion: SKPM results on illuminated devices

We compared the potential distributions of illuminated solar cells with identical active layers equipped with good (MoO_3) and decent (plasma treated ITO) hole extracting contacts. In SKPM studies under illumination and different contact scenarios we found that the solar cells are characterized by two different behaviors when driven at their maximum power point: we found a distinct (donor) bulk limitation for the (good) hole extracting contact with MoO_3/F_4ZnPc and a distinct contact limitation for the contact with plasma treated ITO/F_4ZnPc. Although not surprising in the light of XPS results (see 4.2) indicating the existence of a decomposed F_4ZnPc interlayer of sub-monolayer thickness hindering hole extraction for the plasma treated ITO sample, the strong potential drop limited to the very contact interface area was still unexpected when recalling the decent performance of the thickness-optimized devices (see results in the next section 5.1.2): especially the good

4.4 In-operando SKPM studies on OPV cells with varied hole extracting contacts

FF ($\approx 60\%$) and reasonable V_{oc} ($\approx 640\,\text{mV}$) did not indicate such a distinct difference in the potential distribution between the MoO_3 and plasma treated ITO contact. On the other hand, this result highlights the major role of diffusion in the extraction of photocurrents in organic solar cells. Furthermore we can state that there is a large impact of interface phenomena on SKPM profiles highlighting interface losses in an overproportional way.

Features indicating charge carrier generation at the DA interface could not be detected in both samples.

4.4.3.2 Results 2: Studies on devices under applied bias voltages

Figure 4.14 shows results of measurement series under applied bias voltages on the bilayer solar cells with MoO_3 coated ITO and plasma treated ITO hole extracting contacts. The positive pole is connected to the ITO bottom contact, the Ag top contact is grounded. In the first row single line profiles extracted from the measured SKPM maps (shown here in Appendix) are presented. The relative profiles in the second row are obtained by subtraction of the SC (0 V)-profiles from all other profiles and scaling of these profiles to the bias voltages actually applied to the cells. Overall we find that the trends observed in the studies under illumination are also present here: there is no potential barrier visible at the hole extracting contact of the samples with MoO_3, whereas the samples of plasma treated ITO exhibit potential losses right at the bottom contact. No features are found on the electron extracting Ag top contact, and the major potential drop in both cells occurs in the donor (F_4ZnPc) part of the cell. However, in the MoO_3 sample there is a certain potential drop also in the acceptor (C_{60}) layer.

In the relative profiles of the plasma treated ITO sample we find the potential distributions characterized by a steep potential right at the ITO/F_4ZnPc interface followed by a linear potential drop extending all over the F_4ZnPc layer. All of the voltage applied is consumed in the donor layer, there is no potential drop occurring in the C_{60} acceptor layer. It is probable that the steep potential drop is widened here by a factor of 2-3 because of the limited SKPM resolution and actually constricted to an area of probably $\lesssim 30\,\text{nm}$ from the interface. This is the the upper limit generally found in UPS studies for interface barrier or band bending phenomena (see for example section 4.2). In the following we refer to this steep potential drop as contact

4 Electric potential distribution of F_4ZnPc/C_{60} small molecule organic solar ce

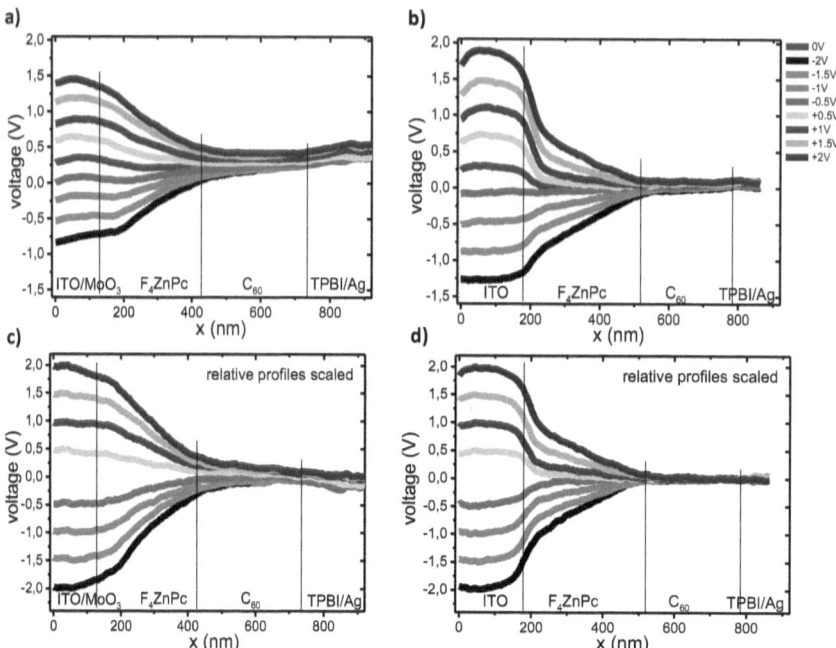

Figure 4.14: SKPM measurement series on cells with ITO/MoO_3 (left column) and plasma treated ITO (right column) hole extracting contact under applied bias voltages from $-2\,V$ to $+2\,V$ in $0.5\,V$ steps. a), b) CPD profiles as-measured in SKPM. Convolution was more pronounced in the measurement on the MoO_3 sample. Applied bias voltages are given on the right hand side. c), d) Relative profiles (SC-profile subtracted) calculated from a), b) and scaled to the actual applied bias voltages. The raw data can be found in Appendix 7.2 and 7.3.

potential barrier. The different contributions to the potential distribution are quantified in table 4.4. We find that the potential barrier is (i) increasing in absolute values and (ii) decreasing relative to the bulk potential drop with increasing applied bias voltages. Whereas the barrier/bulk ratio is ≈ 3 for an applied bias voltage of $\pm 0.5\,V$, it is ≈ 1 for an applied bias voltage of $\pm 2\,V$. We also find that this ratio does not change whether the devices are driven with forward or reverse bias. The contact resistance R_{cont} is approximately constant, so that the voltage drop at the contact increases (almost) linear with the applied voltage. It is probable that this behavior is present also in other organic electronic devices such as organic field effect transistors (OFETs),

4.4 In-operando SKPM studies on OPV cells with varied hole extracting contacts

Applied voltage	Interface (V)	Bulk (V)	Ratio
±0.5V	0.3	0.1	3
±1V	0.55	0.3	1.8
±1.5V	0.75	0.65	1.2
±2V	0.9	0.9	1

Table 4.4: Different potential drops in the relative profiles of the plasma treated ITO sample in figure 4.14. The potential barrier was estimated from the sharp drop on ITO (at $\approx 170\,\mathrm{nm}$) till the transition to the linear regime (at $\approx 215\,\mathrm{nm}$), the bulk potential drop from the linear regime. Values are estimated with a accuracy of $\pm 50\,\mathrm{mV}$.

justifying the assumption of constant contact resistances made for example in the *Transfer Line method* [199] broadly applied for OFET evaluation [200].

In the relative profiles of the MoO_3 sample we find the potential distribution dominated by a strong potential drop in the F_4ZnPc donor layer. Here, no contribution from the hole extracting interface is found, indicating ohmic hole and electron contacts. The potential drops according to a $x^{3/2}$ law from the MoO_3/F_4ZnPc contact interface till about 60 nm from the F_4ZnPc/C_{60} DA interface (according to SCLC theory), where it transitions into a x^{-1} behavior extending into the C_{60} acceptor layer. So opposite to the cells with the plasma treated ITO contact, here a certain voltage drop is also found in C_{60}. We can use this to estimate relative mobilities present in the given F_4ZnPc and C_{60} configuration.

Extraction of F_4ZnPc and C_{60} mobilities

Relative hole (electron) mobilities are extracted here for the F_4ZnPc (C_{60}) layers according to (i) Child, (ii) Poole-Frenkel, (iii) Ohm's and mixed Child/Poole-Frenkel law. For details on these theories it is referred to the fundamentals of charge transport in OSCs discussed in section 2.1.

From the current density found in the IV curve of the solar cell with MoO_3 presented in figure 4.12 we can conclude that for forward bias voltages exceeding $+0.5\,\mathrm{V}$ yields for the charge carrier densities of electrons in C_{60} and holes in F_4ZnPc:

$n_{\text{inj}} \gg n_{\text{intr}}$ and $h_{\text{inj}} \gg h_{\text{intr}}$,

i.e. the charge transport is governed by charge carriers injected from the ohmic hole and electron contacts. In the equilibrium situations probed with SKPM and assuming no (or identical) trap losses for both holes in F_4ZnPc and electrons in C_{60} layers the magnitude of the current densities j in both layers is identical. With

$$j_{F_4ZnPc/C_{60}} = n_{h/e} \cdot e \cdot \mu_{h/e} \cdot F_{F_4ZnPc/C_{60}}$$

and assuming identical relative permittivities for both layers, the law of Child describing SCLCs yields:

$$\frac{j_{F_4ZnPc}}{j_{C_{60}}} = 1 \Rightarrow \frac{\mu_{h,F_4ZnPc}}{\mu_{e,C_{60}}} = \left(\frac{V_{C_{60}}}{V_{F_4ZnPc}}\right)^2, \tag{4.1}$$

where $V_{C_{60}}$ and V_{F_4ZnPc} are the potential drops in the C_{60} and F_4ZnPc layer respectively.

The Poole-Frenkel law uses exponential field-dependencies of zero-field mobilities μ^0 to describe (for example) the field-dependency in hopping charge transport and yields by assuming identical currents j in both layers:

$$\frac{\mu^0_{h,F_4ZnPc}}{\mu^0_{e,C_{60}}} = \frac{F_{C_{60}}}{F_{F_4ZnPc}} \exp\left[\frac{e^{3/2}}{2k_BT} \frac{1}{\sqrt{\pi \varepsilon_0 \varepsilon_r}} \left(\sqrt{F_{C_{60}}} - \sqrt{F_{F_4ZnPc}}\right)\right], \tag{4.2}$$

with the electric fields $F_{C_{60}}$ and F_{F_4ZnPc} in the C_{60} and F_4ZnPc layer respectively. We assume a linear potential evolution (constant fields) and a relative permittivity of $\varepsilon_r = 3.5$ for both F_4ZnPc and C_{60} here.

For comparison also Ohm's law is consulted, which yields:

$$\frac{\mu_{h,F_4ZnPc}}{\mu_{e,C_{60}}} = \frac{F_{C_{60}}}{F_{F_4ZnPc}}. \tag{4.3}$$

4.4 In-operando SKPM studies on OPV cells with varied hole extracting contacts

Voltage	j $\left(\frac{mA}{cm^2}\right)$	V_{F_4ZnPc} (V)	$V_{C_{60}}$ (V)	μ_{rel} Child	μ_{rel} Poole-Frenkel	μ_{rel} Ohm	μ_{rel} C+P-F
+0.6 V	0.003	0.5	0.07	0.02	0.05	0.14	0.01
+0.8 V	0.010	0.65	0.1	0.02	0.07	0.15	0.01
+1 V	0.029	0.75	0.15	0.04	0.1	0.2	0.02
+1.5 V	0.191	1.2	0.15	0.02	0.04	0.13	0.005
+2 V	0.645	1.5	0.25	0.03	0.06	0.17	0.01

Table 4.5: Relative F_4ZnPc hole mobilities $\mu_{rel} = \frac{\mu_{h,F_4ZnPc}}{\mu_{e,C_{60}}}$ in units of C_{60} electron mobilities calculated from the potential drops V_{F_4ZnPc} and $V_{C_{60}}$. The potential drops are determined from relative profiles of the bilayer cells with ohmic MoO_3/F_4ZnPc contact presented in figure 4.14. The current density j is taken from the IV curve shown in figure 4.12. Mobilities are calculated using SCLC (Child), hopping (Poole-Frenkel), ohmic (Ohm) and hopping SCLC transport (Child+Poole-Frenkel).

We extracted the potential drops for both layers from the relative (and scaled) profiles of the MoO_3 sample in figure 4.14 and calculated relative mobilities according to the mentioned models. The results of F_4ZnPc hole mobilities given in units of C_{60} electron mobilities are presented in table 4.5. Besides the discussed expressions the mobilities were also calculated assuming hopping transport in a SC limited environment by multiplying the exponential factor of equation 4.2 to the value obtained by equation 4.1. Overall we find good agreement for the values extracted by the respective models for all bias voltages.

Opitz et al. found hole (electron) mobilities of $\mu_{h,CuPc} \approx 10^{-5}$ cm^2(Vs)$^{-1}$ ($\mu_{e,C_{60}} \approx 10^{-1}$ cm^2(Vs)$^{-1}$) for CuPc (C_{60}) in SCLC diodes of 200 nm thickness [6] [133]. However, we found significantly higher (relative) values for the F_4ZnPc mobility here. If we assume the C_{60} electron mobility to be $\mu_e \approx 10^{-1}$ cm^2(Vs)$^{-1}$ as given in literature, the (zero-field) hole mobilities we find for F_4ZnPc are

$$\mu^{SCLC}_{h,F_4ZnPc} \approx (3 \pm 1) \cdot 10^{-3}\, cm^2(Vs)^{-1},$$

---

[6] Lacking literature values, we assume $\mu_{h,F_4ZnPc} \approx \mu_{h,CuPc}$ here (see section 3.2.1).

$$\mu^{\text{hopping}}{}_{\text{h},\text{F}_4\text{ZnPc}} \approx (7 \pm 3) \cdot 10^{-3}\,\text{cm}^2(\text{Vs})^{-1},$$

$$\mu^{\text{SCLC+P-F}}{}_{\text{h},\text{F}_4\text{ZnPc}} \approx (1 \pm 0.5) \cdot 10^{-3}\,\text{cm}^2(\text{Vs})^{-1}$$

according to the SCLC/Poole-Frenkel hopping /mixed SCLC-hopping model. So the mobilities we found here are about two orders of magnitude higher than the ones found by Opitz et al. in SCLC structure. From a field effect transistor structure they reported values of $\mu_{\text{h,CuPc}} = 1.8 \cdot 10^{-3}\,\text{cm}^2(\text{Vs})^{-1}$ [132], which are in good agreement with our values.

Of course, the discussion held could be applied in the same manner to the electron mobility of C_{60}, which would be found to be about two orders of magnitude lower than the literature values. Also a mixture of both higher F_4ZnPc hole and lower C_{60} electron mobility is thinkable.

Conclusion: SKPM results on devices under applied bias voltages

We operated bilayer organic solar cells with good (MoO_3) and decent (plasma treated ITO) hole extracting contacts by applying bias voltages under dark conditions and studied their potential distribution with SKPM. The C_{60}/TPBi/Ag electron extracting top contact was found to exhibit ohmic behavior both for charge extraction and injection, as no potential drops were detected here in any configuration.

For the cells with (decent) ITO hole extracting contacts, we found that the applied bias voltage drops both in the F_4ZnPc donor bulk and at the very ITO/F_4ZnPc interface, indicating a potential barrier located here. The C_{60} acceptor layer was found field-free for the bias voltage range probed ($V_{\text{app}} \leq \pm 2\,\text{V}$). The potential barrier at the ITO/F_4ZnPc interface can be approximated assuming a constant contact resistance R_{cont}. The potential drop in the F_4ZnPc donor bulk was linear.

For the cells with (good) MoO_3 coated ITO hole extracting contacts we found ohmic contact properties for both electron and hole contacts. The applied bias voltage drops in the active layer bulk only, with no features visible in the contact region. This fact is exploited to study the bulk properties of F_4ZnPc and C_{60}. Applying models describing SCLC (Child), hopping (Poole-Frenkel), ohmic (Ohm) and hopping SCLC (Child + Poole-Frenkel) transport to potential distributions of solar cells in the high injection regime ($V_{app} > +0.5\,V$), we extracted relative hole (electron) mobilities in F_4ZnPc (C_{60}). We found a good agreement over the entire voltage range in the framework of the respective theories. Thereby relative mobilities calculated according to Poole-Frenkel law were about a factor 2 higher than mobilities calculated according to Child law. Deviating from mobility ratios of 10^{-4} for μ^{SCLC}_{h,F_4ZnPc} and $\mu^{SCLC}_{e,C_{60}}$ reported in literature, here we found a ratio of $3 \cdot 10^{-2}$ when SCLC theory was applied. For hopping transport under SCLC conditions a ratio of 10^{-2} was found. Probably there is a certain underestimation (overestimation) of the voltage drops extracted from the F_4ZnPc (C_{60}) layers because of SKPM convolution, accounting for a part of the deviation from literature values. However, the ratio of 10^{-4} reported for μ^{SCLC}_{h,F_4ZnPc} and $\mu^{SCLC}_{e,C_{60}}$ would require potential drops of $1.98\,V$ in the F_4ZnPc and $0.02\,V$ in the C_{60} layer here (for $V_{app} = +2\,V$). So we think the deviation can not be attributed to convolution only, but rather stems from the differences in the sample geometry.

4.4.4 Discussion: Prospects and limits for in-operando SKPM studies

The aim of this study was the detailed mapping and understanding of the electric potential evolution in organic solar cells based on in-operando SKPM characterization. From findings in literature and previous studies performed at iL we were aware of the challenges that are posed on the road to qualitative understanding and quantitative evaluation of in-operando SKPM results. The most prominent are (i) the generation of meaningful SC profiles, (ii) the need of profound knowledge on device performance, materials and especially interfaces to understand SKPM results qualitatively and (iii) the description of SKPM results in the framework of drift-diffusion theories.

We faced these challenges in a twofold manner: on the cross section preparational side we advanced our set of methods by state-of-the-art FIB preparation techniques to head for minimization of influencing the electric properties of the sample surface. On the device side we switched from solution processed to vacuum processed small molecule solar cells based on the material system

F_4ZnPc/C_{60}. With the characterization methods available in clustertool, interface phenomena in vacuum processed devices were accessed with high definition and the material system was approached under different perspectives within the analytics group [37, 152, 175, 201], so that proper knowledge on the physics of the entire device was achieved (see for example next section). We correlated the potential distribution from SKPM studies with findings from complementary characterization techniques such as XPS/UPS and KP as well as with device performance. Different FIB ions were used to achieve samples of high quality. Still we found that some of the questions addressed remained open. Here we first give an overview on the main issues in in-operando SKPM studies before we come to the prospects of this method in the field of organic electronics.

It is still unclear what prevents the extraction of meaningful SC potential profiles with SKPM. As well-known from Ga FIB preparation, also the FIB preparation using Ne and He ions does not lead to profiles that agree with results from layer-by-layer characterization with KP and UPS.

When regarding the prominent role of the DA interface OPV, it is difficult to understand why the measurements of illuminated cells do not exhibit any specific feature at this interface between F_4ZnPc and C_{60}. On the one hand, it should not be excluded that preparation induced strong surface doping is responsible for this. On the other hand, in the light of the findings concerning bulk and contact limitation found in great detail, the reason of preparational damage is very improbable. In fact, the impossibility to address this central issue puts an end to all efforts towards meaningful correlation of in-operando SKPM results with drift-diffusion theory.

When considering the results under applied bias voltages presented here as well as in all studies reported so far one notices that there is a total symmetry in the relative profiles of positive and negative applied bias voltages, i.e. the profiles from samples under $+V_{app}$ are equal to the the profiles under $-V_{app}$ mirrored on the horizontal 0 V-line. When considering the results under illumination we find that inducing OC conditions by applying the bias voltage $V = V_{oc}$ leads to the same (relative) potential profile as disconnecting one of the contacts. At first glance, this could be interpreted as a validation of the method (as indeed done by Chen et al. [196]). However, as discussed in section 3.1.2.1 the relative potential profiles represent the net effect of the

4.4 In-operando SKPM studies on OPV cells with varied hole extracting contacts

bias voltage/illumination on the device and it should not taken for granted that this effect manifests itself in the same way when (for example) the solar cell is driven under forward and reverse bias, because the underlying mechanisms in the different material phases (conduction) and at the interfaces (injection/extraction) of the cell vary strongly. This is discussed here on the example of the biased bilayer solar cells with (good) MoO_3 coated ITO contacts. In these measurements we could confirm literature results from Opitz et al. who found the hole mobility of F_4ZnPc lower by several orders of magnitude than the electron mobility of C_{60} [133]. Under applied forward bias and sufficient hole (electron) injection into F_4ZnPc (C_{60}) the gap in mobility induces the potential profiles observed characterized by a strong (slight) potential drop over the F_4ZnPc (C_{60}) layer. However, Opitz et al. found also that the hole mobility of C_{60} is about one magnitude smaller than the electron mobility of F_4ZnPc. So in reverse bias, with hole (electron) injection into C_{60} (F_4ZnPc), this should lead to a significantly higher potential drop in the C_{60} layer in comparison to the forward biased case, which is not observed. The relative shift of the contacts regardless of its origin is the only parameter dictating the relative potential profiles observed in the device.

An additional issue limiting the quantitative power of in-operando SKPM is cantilever-sample convolution. As demonstrated in the evaluation of our measurements, the reduced absolute shift of the contacts can be corrected easily. Same applies to the extraction of electric field strengths, which become underestimated by convolution, but still exhibit the right trends. However, this does not apply to the shape of potential drops at the contacts, with dramatic impacts on the extraction of charge carrier densities. As becomes evident from the bowl-shape, the as-measured profiles do not lead to meaningful results here. However, also fitting does not yield meaningful results if the actual shape of the potential drop is not known (for example from complementary methods): typical shapes obtained in our measurements for biased cells with (decent) ITO contacts are fitted well by the inverse of a hyperbolic cosine function $\pm \cosh^{-1}(x)$ (as proposed by Mankel et al. for interfaces in SC [202]) or a parable $\pm x^2$ in proximity of the contact. However, due to convolution an exponential potential drop at the contact would also lead to the same SKPM profile. The 2nd derivation ($\sim$ charge carrier density) of the discussed fitting functions does not only yield deviating values, but even different signs, proving the inapplicability of this approach.

The in-operando SKPM results of distinctively differing behavior induced by the different contacts for identical stacks otherwise discussed here demonstrate that our approach of FIB exposing the solar cell cross section and in-situ characterization with SKPM yields true information on the device physics in line with literature reports and is not governed by artifacts induced by sample preparation. Furthermore we found that interface properties at the contacts contribute in an overproportional manner. Required are active layers that are sufficiently thick to separate contact and bulk contributions. This makes in-operando SKPM characterization an appropriate tool for the investigation of solution processed devices because of mainly two reasons: (i) Solution processed devices are typically realized using thicker active layers than found in vacuum processed devices, so that the need of thick active layers is fulfilled here more naturally. (ii) Because its high sensibility towards interface phenomena at the contacts, in-operando SKPM could act as a screening method in contact engineering of solution processed OLEDs and OPV devices using new materials with unexplored properties regarding interface formation. Solution processed devices are significantly less accessible for established characterization methods than vacuum processed devices, this regarding to methods both addressing bulk and interface phenomena. Whereas for vacuum processed devices both top and bottom contact formation as well as organic heterojunctions can be explored in large detail by means of a layer-by-layer characterization with UPS, in solution processed devices this possibility is restricted. The same applies to the potential evolution within the active layers of operating devices, which can be explored for (simple) vacuum processed devices by means of an IV measurement series of devices with varying layer thickness [203]. This is also possible for solution processed devices, but significantly larger experimental efforts are required to scan the entire range of thicknesses here. Also in the field of degradation studies important issues concerning device physics could be addressed by in-operando SKPM, as demonstrated successfully by Weigel et al. [174].

5 Structure-function relationship in F_4ZnPc/C_{60} solar cells

In this chapter we present studies of $F_4ZnPc:C_{60}$ solar cells prepared under controlled variation of the fabrication conditions. The performance of the devices is discussed in the light of findings from electronic and structural investigations with TEM, AFM and XPS.

Here, the fabrication of OPV devices was performed with the objective of a deeper understanding of fundamental processes in modern OPV. Thus the focus was on simple device architectures close to architectures of samples we characterized analytically. On the one hand, reasonable efficiencies had to be achieved to ensure relevance of the analytical results. On the other hand, we did not perform a full stack optimization. The focus was on analytical studies on the OPV systems. For this, the OLED1 chamber was equipped with solar cell materials and a substrate heating unit, allowing for state-of-the-art organic solar cell preparation. More on the experimental details can be found in section 3.2.3.

First we present results on conventional (non-inverted) bilayer solar cells with different hole extracting contacts. Combined AFM/XPS studies on these cells address the issue of growth and coverage. Afterwards, results are presented on BHJ solar cells processed in both inverted and conventional layer stacking with varied substrate temperature during active layer deposition. The results are discussed in the light of findings from analytical TEM studies revealing morphology and crystallinity of the BHJ on the nanoscale.

All IV curves presented were derived from 2-point IV measurements. The solar cell results are discussed using the concepts of electrical characterization introduced in section 2.2.1. For details on the materials and preparation see section 3.2.1 and 3.2.2. The active area of all solar cells discussed here was $4\,\mathrm{mm}^2$.

5.1 Bilayer solar cells with varied hole extracting contact

In this section we discuss results on $F_4ZnPc:C_{60}$ bilayer solar cells with varied hole extracting contact: ITO coated with MoO_3 and NiO_x as well as O_2 plasma treated ITO are used for efficient hole extraction. The performance of the device is discussed in the light of the findings from the previous chapter. The coverage and roughness of the active materials deposited on substrate and underlying organic layers have a significant impact on the choice of optimal layer thicknesses in the stack. Here, based on in-situ AFM and XPS studies we comment on growth and coverage in the solar cell stack. The fundamentals of AFM and XPS are discussed in section 3.1.1 and 3.1.7. First, an introduction in this field is given.

5.1.1 State of the art

As discussed in section 2.2 the fundamental difference of organic to inorganic solar cells is the mechanism of charge carrier separation after light absorption. The high polarizability of the organic matrix necessitates the need of chemical energy provided at the donor/acceptor (DA) interface for efficient exciton separation. Tang was the first who exploited the principle of the organic-organic DA heterojunction systematically and achieved organic solar cells with PCEs exceeding 1% [8]. His bilayer stack architecture of two organic thin films with distinct D/A electronic properties is applied also in the bilayer cells presented here. Whereas he used a perylene derivative as electron acceptor, fullerenes became standard in the meantime [204]. However, the donor molecule F_4ZnPc used here is similar to CuPc which was used by Tang: both are phtalocyanines with a central metal atom. The fluorinated ZnPc is tailored for OPV applications, providing higher V_{oc} because of its beneficial energy alignment with C_{60} [17,33]. First optoelectronic experiments on the effects of fluorination of phtalocyanines go back to the early 2000s [205], but its application in OPV is still a matter of research [160,206]. One drawback of the phtalocyanines for OPV applications is their poor conductivity. To account for this, often very thin phtalocyanine layers of < 20 nm combined with C_{60} layers of $\gtrsim$ 50 nm thickness are applied [207, 208]. Here, comparability to the devices analyzed in the previous chapter was of special interest, so layer thicknesses of 30 nm were chosen both for C_{60} and F_4ZnPc. Tang stated the importance of ohmic contacts for charge carrier extraction already in his pioneering work. In the 2010s, the application of buffer layers in OPV became a main focus of attention: the blocking and charge extraction properties of interfacial materials were widely discussed based both on experiments [209, 210] and theory [211]. Transparent transition metal oxides (TMOs) proved to provide excellent hole extracting/injecting contacts for many applications in organic optoelectronics [142, 212]. Because of their very high work functions they ensure hole extraction also in combination with low HOMO level materials such as F_4ZnPc. Here, MoO_3 is used as hole extracting contact. However, the electron blocking capabilities of MoO_3 are limited [142, 144]. Nickel oxide (NiO_x) proved to be a good alternative closing this gap. Many studies report on its strong electron and exciton blocking properties and excellent suitability as hole extracting contact in solution processed OPV [149, 213, 214]. Schulz et al. demonstrated that the NiO_x electron/exciton blocking properties can be combined with the hole extracting properties of MoO_3 by applying a 2 nm thin layer of MoO_3 on the NiO_x coated ITO substrate before the deposition of the BHJ [144].

AFM studies are widely used in research on organic solar cells for the monitoring of coverage [215] and orientation [216–218] of organic molecules typically on electrodes. Studies on heating or substrate induced roughness are mandatory for the identification of loss paths in the device [22, 134, 136]. In research on organic electronics, only very few AFM studies are performed in UHV [219, 220]. In inorganic surface science in-situ AFM applied in UHV as well as XPS are standard tools today. However, studies correlating in-situ characterization of XPS and AFM as presented here are rare because of the need of an integrated UHV setup [221].

5.1.2 Solar cell results on F_4ZnPc/C_{60} bilayer devices

We fabricated F_4ZnPc/C_{60} bilayer solar cells in conventional solar cell architecture. The active photovoltaic layer consisted of 30 nm F_4ZnPc and 30 nm

5 Structure-function relationship in F_4ZnPc/C_{60} solar cells

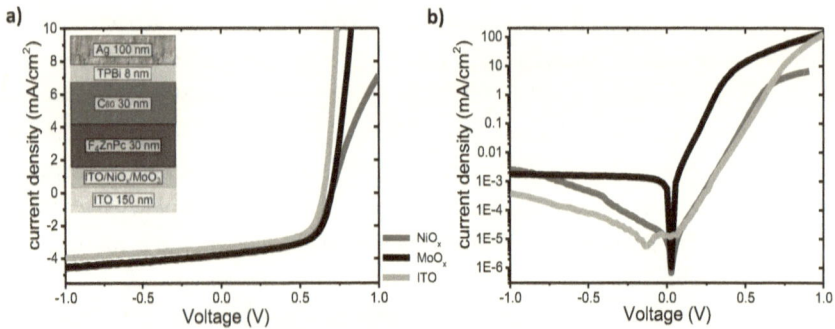

Figure 5.1: IV characteristics of F_4ZnPc/C_{60} OPV bilayer cells in conventional architecture (best cell of the batch). MoO_3 (black), NiO_x (green) and O_2 plasma treated ITO (orange) were used as hole extracting contacts, TPBi as electron extracting contact. a) IV characteristics under AM1.5 illumination. Inset: the used solar cell stack. b) IV characteristics under dark conditions.

C_{60}. The ITO substrate underwent different surface treatments: we used MoO_3, NiO_x and O_2 plasma treated ITO to obtain efficient hole extracting ITO contacts. All devices presented were characterized in a glovebox under N_2 atmosphere.

Figure 5.1 shows the IV curves of $F_4ZnPc:C_{60}$ bilayer champion devices with MoO_3 (black), NiO_x (green) and O_2 plasma treated ITO (orange) as hole extracting contacts under AM1.5 illumination and in the dark. The performance of the devices is compared in table 5.1: the best devices have PCEs of $1.6 \pm 0.1\%$. This performance is in line with results reported in literature for similar and identical devices [160, 222].

Comparing the absolute PCE of the three different cells, we find MoO_3 to be the best performing stack combination. As shown in UPS interface experiments presented in section 4.2, bilayer cells based on a MoO_3/F_4ZnPc hole extracting interface exhibit a strong band bending of $eV_{bb} = 0.6\,eV$ extending in a significant part of the donor layer and high built-in voltages of up to $V_{bi} = 2.0\,eV$. Whereas the former supports hole transport in the F_4ZnPc donor layer, the latter allows for high V_{oc} (see section 2.2.4 for the role of V_{bi} in OPV). As known from literature, the electron blocking properties of MoO_3 are limited [142, 144], but are of secondary importance in the bilayer architecture here. Probably because of this, the V_{oc} achieved in the MoO_3

5.1 Bilayer solar cells with varied hole extracting contact

	MoO_3	$ITO\ (O_2)$	NiO_x
V_{oc} (mV)	692 ± 12	636 ± 14	707 ± 10
j_{sc} (mA/cm^2)	3.8 ± 0.1	3.4 ± 0.1	3.7 ± 0.1
FF (%)	59.4 ± 0.7	60 ± 1.4	55.2 ± 0.5
Sat	1.19	1.19	1.22
η (%)	1.6 ± 0.1	1.2 ± 0.1	1.5 ± 0.1

Table 5.1: Characteristics of bilayer solar cells with varied hole extracting contact. The values represent the average of the four best cells each. Cells with MoO_3 and NiO_x coated ITO both reach high V_{oc} as well as high j_{sc}, outperforming the O_2 plasma treated ITO substrate. The latter exhibits the highest FF, probably because of the lower current. The saturation is extracted from the best cells shown in figure 5.1: all devices exhibit decent saturation of about Sat ~ 1.2.

devices was almost as high as in the devices with NiO_x contacts.

As expected from its excellent electron blocking properties, NiO_x led to the highest V_{oc} obtained here. However, also very high short circuit currents were achieved, in line with results reported from Manders et al. [149]. They compared solution-processed OPV with PEDOT:PSS and NiO_x hole extracting contacts and found the superiority of the NiO_x samples based first of all on increased j_{sc} rather than on significantly elevated V_{oc}, as expected. The lower fill factor compared to the other anode materials is probably related to a higher series resistance of the NiO_x cells, indicated by the low forward currents.

Bilayer devices with O_2 plasma treated ITO anode exhibit the lowest PCE, as expected when recalling the results from the previous chapter (see 4.2). The lower band bending in the donor region following the ITO/F_4ZnPc contact ($eV_{bb} = 0.45\,eV$) compared to the MoO_3 sample could be responsible for the lower j_{sc}. The lower V_{oc} could be caused by the higher hole extraction barrier ($\Delta_{holes}^{ITO} = 0.5\,eV$ instead of $\Delta_{holes}^{MoO3} = 0.3\,eV$) and the lower built-in voltage of the ITO devices. However, the decomposed F_4ZnPc right at the interface with the O_2 plasma treated ITO seems to be of minor importance for the device performance. Although also this could lead to lower V_{oc} and j_{sc}, the impact of the decomposed layer on device performance is surprisingly low.

5 Structure-function relationship in F_4ZnPc/C_{60} solar cells

Conclusion: bilayer solar cell results

We developed simple OPV stacks for efficient vacuum processed small molecule F_4ZnPc/C_{60} solar cells in conventional bilayer architecture. The performance of solar cells fabricated on MoO_3 coated ITO (PCE = $1.6 \pm 0.1\%$) keep up with the latest results reported in literature.

The beneficial effect of electron/exciton blocking is less important in the bilayer device structure used here than it is in solution-processed BHJ cells, characterized by the direct contact of BHJ and hole extracting contact. Likely because of this the devices on NiO_x exhibited slightly lower efficiencies of 1.5% compared with the devices on MoO_3.

As expected, the devices fabricated on O_2 plasma treated ITO exhibited the lowest PCE. However, a lowering of PCE by about 20% with respect to the MoO_3 substrate seems moderate when recalling the findings from the previous section.

Results from step-by-step AFM growth studies on the bilayer stack presented in the next section demonstrate the distinctive smoothening effect of the TPBi capping layer. We think that besides the hole blocking effect of the closed TPBi layer also this smoothening contributes significantly to the good performance achieved for all devices.

5.1.3 In-situ monitoring of bilayer thin film growth

Vacuum processed small molecule organic solar cells are characterized by very thin active layers of about 50 to 100 nm. Because of this already small variations in film thickness and roughness can have a significant impact on the device performance. Be it because of modifications in the potential distribution within the device (establishing field and thus charge transport inhomogeneities) or because of direct contact between the active layer and the electrode (enhancing recombination losses), roughness and thickness variations can induce or enhance existing loss paths in the device. Here, we present studies addressing the issue of organic thin film growth from a morphological point of view. A series of in-situ AFM measurements on the F_4ZnPc/C_{60} bilayer stack grown on MoO_3 and characterized in the previous section 5.1.2 provides topographical insights. XPS data comparing the F_4ZnPc growth mode of a MoO_3 coated ITO and an O_2 plasma treated ITO substrate

5.1 Bilayer solar cells with varied hole extracting contact

complete the picture.

Experimental details

The AFM measurements were performed parallel to (and on the same samples as) the KP studies on the step-by-step grown bilayer stack presented in section 4.1. The samples were prepared at clustertool in the organic (OLED1) and metal preparation chambers and transferred for characterization to the UHV-SPM chamber without breaking the UHV. Here, the AFM measurements at certain stack positions were performed. Afterwards, deposition was continued and the next layer applied.

We want to emphasize that the microstructure obtained from AFM characterization is given by the convolution of the surface with the cantilever tip. Therefore, especially when comparing AFM data quantitatively, one should ensure the use of identical cantilevers. Here, we used cantilevers of the same model: for the measurements at the positions 2, 4, 6, 7 and 8 same cantilever (A) was used, for measurements at position 3 and 5 cantilever B was used. Position 1 was measured with a third cantilever and at ambient. For further details on methods and instruments involved see section 3.

5.1.3.1 Growth and coverage studied with AFM

In figure 5.2 the AFM micrographs taken at different bilayer stack positions are presented. Graph 1 shows the AFM image of a (O_2 plasma treated) ITO surface. The ITO-typical structure is present here: flakes with a size of some hundred nm mantled with small spheres of some tens of nm in size. After depositing 10 nm of MoO_3 (graph 2) the small spheres are smoothed, but the flake-like structure is still dominant. Same applies after the deposition of 5 nm and even 30 nm of F_4ZnPc (graph 3 and 4). However, already in graph 3 there is a clear change on the scale of < 100 nm compared to graph 2: a new growth mode is emerging, characterized by small nanostructures in the order of some ten nm in lateral and a few nm in z direction [1]. These grow bigger

---

[1] We want to note here that graph 3 was measured with a different cantilever (B) than graph 2 (A). The emergence of the nanostructure could thus also be caused by a smaller tip radius of cantilever B. However, in graph 8 (recorded after graph 2) nanostructures on the same order of the ones in graph 3 are resolved with cantilever A. This, in combination with control measurements, showed us that there is indeed a change in the surface topography here. Same applies to graph 4 and 5.

5 Structure-function relationship in F_4ZnPc/C_{60} solar cells

with further deposition of F_4ZnPc (graph 4). Both observations, the retained flake structure on the bigger and the emerging structure on the smaller scale, indicate a rather uniform F_4ZnPc growth on the MoO_3 coated ITO substrate for the first few monolayer, followed by the emergence of small molecular islands (Stranski-Krastanow growth). This is in line with a growth mode estimation extracted from XPS data (figure 5.3) that indicates a full coverage of the MoO_3 substrate after the first 2 nm. After depositing 5 nm C_{60} on F_4ZnPc (graph 5), we see a similar behavior as in graph 3: a new structure emerges on the scale of some ten nm in lateral, which is growing bigger with further deposition (visible in graph 6). With XPS it was shown that C_{60} covers a F_4ZnPc layer already after some monolayer [33], so that also here the assumption of Stranski-Krastanow growth (closed monolayer(s) followed by island growth) is reasonable. After the deposition of 30 nm C_{60} (graph 6), the topography is dominated by C_{60} islands, apparently expanding from the smaller islands visible in graph 5. Most of them are coalesced, building a rather homogeneous layer of a few nm roughness only. However, some of the islands grow significantly bigger and exceed the average roughness by more than 10 nm. This increased roughness is smoothed by the following TPBi capping layer. As visible in graph 7, the 8 nm thin TPBi layer covers the underlying C_{60} layer entirely. This underlines the excellent applicability of TPBi as electron capping layer in OPV: it prevents direct contact between active layer and metal contact, enabling high V_{oc} by suppressing hole recombination at the electron contact. Furthermore, the reduced roughness (rms reduction is > 50%) ensures high lateral homogeneity of the electric field within the device. The 100 nm thick silver top electrode growths smoothly on TPBi (graph 8).

Roughness and interfacial area

In table 5.2 the roughness parameter extracted from the AFM measurements are shown. The root mean square (rms) as well as the actual surface area (area) and the ratio of actual and protected area (ratio) are given. The actual surface area represents the interface area between sample surface and vacuum as determined by AFM. Roughness and elevations in topography increase the actual surface area, which is an instructive measure for the actual

5.1 Bilayer solar cells with varied hole extracting contact

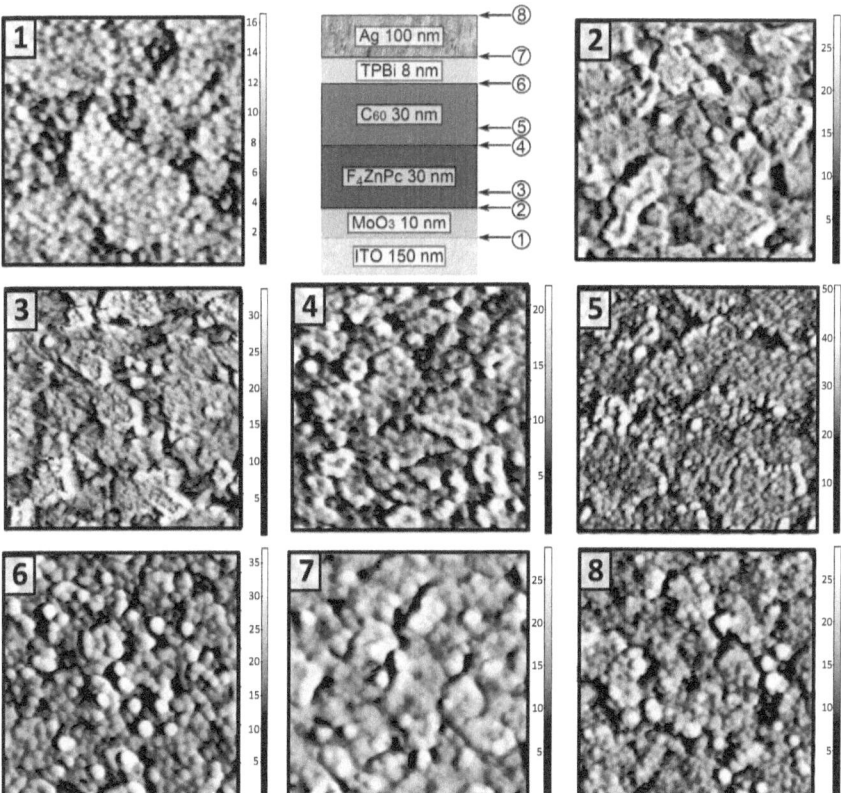

Figure 5.2: AFM micrographs taken at distinctive positions in the stack. All micrographs have a size of $1.3 \cdot 1.3\,\mu m^2$. The arrows at the stack mark the measurement positions. All layers are reasonable smooth. The typical ITO flake-like structure is visible also after the deposition of F_4ZnPc. The TPBi capping layer smooths the surface significantly, preventing direct contact between the Ag top contact and the C_{60} acceptor.

5 Structure-function relationship in F_4ZnPc/C_{60} solar cells

	ITO	MoO_3	F_4ZnPc^1	F_4ZnPc^2	C_{60}^1	C_{60}^2	TPBi	A
rms (nm)	2.2	3.8	3.6	3.4	4.3	4.3	2.8	3.
area (μm^2)	1.70	1.73	1.74	1.72	1.78	1.75	1.70	1.7
ratio (%)	0.7	2.4	3.0	1.8	5.3	3.6	0.6	1.

Table 5.2: Roughness parameter extracted from the AFM micrographs shown in figure 5.2. The actual surface area (area) is given by the interface area between the sample surface and air. Ratio is given by the difference of the actual and the projected area in percentage. The projected area is $1.69\,\mu m^2$ for all samples. F_4ZnPc^1 (F_4ZnPc^2) and C_{60}^1 (C_{60}^2) refer to the AFM measurement after deposition of 5 nm (30 nm) F_4ZnPc and C_{60} respectively.

interface area between the given and the following layer in the stack. The ratio between actual and projected area quantifies the *increase* in interfacial area induced by the given surface roughness with respect to a totally flat interface (given here in percentage). Thus it is a meaningful measure when it comes to the discussion of interface phenomena such as interfacial recombination. Overall, we find a rather smooth layer growth for all characterized positions in the stack. There is a constant rms value of 3.6 ± 0.2 nm for the first three layers applied. Same applies to the actual surface area. This confirms the interpretation of the AFM micrographs: if applying F_4ZnPc on MoO_3 coated ITO with given thicknesses, the underlying ITO structure is more or less retained. Gommans et al. published AFM and SEM micrographs of thin films of SubPc on ITO which exhibited very similar behaviour [215]. Only with the application of C_{60} the surface roughness as well as the actual surface area increase significantly. As discussed above, the 8 nm thin TPBi capping layer decreases the surface roughness significantly. The roughness induced increase of the interfacial area to the silver top contact is 0.6% only, whereas it is 3.6% for the C_{60} layer.

5.1.3.2 F_4ZnPc growth studied with XPS

Figure 5.3 shows a comparison of the F_4ZnPc coverage on a MoO_3 coated ITO substrate and an O_2 plasma treated ITO substrate determined with XPS. Integrated peak intensities of characteristic XPS resonances stemming

5.1 Bilayer solar cells with varied hole extracting contact

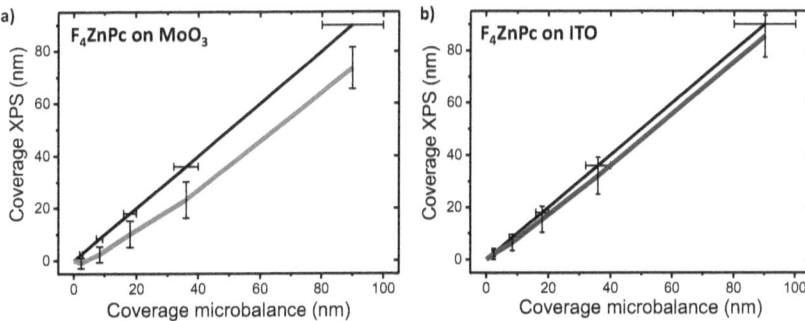

Figure 5.3: The F_4ZnPc coverage of the a) MoO_3 and b) O_2 plasma treated ITO substrate obtained from the damping of a) Mo3d and b) In3d peak intensity is plotted versus the coverage as given from the quartz microbalance. Whereas the XPS coverage of the MoO_3 substrate exhibits a slight kink for sub-monolayer thicknesses, the XPS coverage of the ITO grows strictly linear with the value of the microbalance.

from the substrate are plotted versus the layer thicknesses determined with the quartz microbalance in the preparation chamber. The XPS signal of the substrate is attenuated with increasing layer thickness and an effective layer thickness is calculated using a Beer-Lambert approach (see section 3.1.7 for details). With this the growth mode and sticking coefficient of the specific material-substrate combination can be studied: whereas a change in growth mode leads to a different slope in the diagram, a low sticking coefficient leads to lower y-axis intercepts. The reason for this is the less effective sticking of the material to the substrate compared to the organic-organic sticking to the already coated quartz microbalance. Here, (i) Mo3d and (ii) In3d peak intensities from the XPS measurements presented in figure 4.5 are used to analyze F_4ZnPc growth on (i) MoO_3 coated ITO and (ii) O_2 plasma treated ITO.

From the XPS coverage of F_4ZnPc on MoO_3 we find that the sticking coefficient for this material combination is limiting the controlled deposition of sub-monolayer films. There is a kink for the 2 Å layer thickness indicating reduced sticking of F_4ZnPc at the very MoO_3 surface. However, this effect is reasonable small here: from the 8 Å layer on F_4ZnPc grows in a linear fashion, i.e. the XPS and the microbalance coverage overlap. As the slope does not undergo a significant shift here, there is no indication of a distinctive island

117

(Volmer-Weber) growth in the first few monolayer.

In the case of the O_2 plasma treated ITO surface all arriving molecules stick to the surface and there is a linear behaviour in the XPS coverage from the first monolayer on. This is in line with our results presented in section 4.2, where we found the ITO chemically activated by the plasma treatment. Because of this high reactivity the sticking to the surface is very strong. Also here, there is no indication of of Volmer-Weber growth, but closed layers are built from the first monolayer on.

5.1.3.3 Conclusion: in-situ monitoring of bilayer stack growth

With a combined AFM/XPS study we monitored the growth of a F_4ZnPc/C_{60} bilayer stack in-situ at eight stack positions. From XPS data we found that the sticking coefficient for F_4ZnPc on MoO_3 coated ITO is < 1, limiting growth in the (nominal) sub-monolayer regime. Furthermore it was demonstrated that F_4ZnPc builds closed layers on this substrate already after few (nominal) monolayer. Same applies for C_{60} deposited on F_4ZnPc. For both interfaces (MoO_3/F_4ZnPc and F_4ZnPc/C_{60}), results from AFM micrographs indicate a Stranski-Krastanow growth: after closed layers are established, islands with a lateral expansion of some tens of nm emerge. Concerning the stack design of conventional BHJ solar cells we can conclude that a pristine underlayer of 5 nm F_4ZnPc is sufficient to prevent direct contact between the BHJ and the hole extracting contact. In terms of roughness all surfaces grow reasonably smooth. Still, a significant increase in roughness was found after the deposition of C_{60}. This was countered by the smoothening effect of the TPBi capping layer: 8 nm TPBi lowered the surface roughness of the stack by more than 50%. With this, the interfacial area between the organics and the silver contact is practically that of a plain interface, emphasizing the good applicability of TPBi as electron capping layer.

5.2 C_{60} crystallinity dictates device efficiency in $F_4ZnPc:C_{60}$ BHJ solar cells

In this section we present investigations on $F_4ZnPc:C_{60}$ BHJ solar cells processed on different substrates and at different substrate temperatures.

5.2 C_{60} crystallinity dictates device efficiency in $F_4ZnPc:C_{60}$ BHJ solar cells

The impact of these process conditions on the solar cell performance is studied in IV measurements. Resulting trends are correlated to studies on the nanostructure of the active layer blend performed with TEM and AFM. With analytical TEM we reveal morphology and crystallinity of the BHJ blends, AFM measurements are used to cross-check the TEM results. By correlating device and TEM data we are able to develop a general model providing design criteria for efficient vacuum processed organic solar cells.

5.2.1 State of the art

Charge carrier harvesting in modern organic solar cell devices is based on the highly efficient exciton splitting at the organic donor-acceptor interface [223]. A huge boost in organic solar cell research came in the mid-1990s, when the research group around Alan J. Heeger introduced the concept of the bulk heterojunction (BHJ): they proved that a network of DA molecules interpenetrating on the nanoscale (BHJ) can lead to significantly higher PCEs when compared to plane bilayer heterostructures [48]. More DA interface per volume unit is provided, leading to enhanced harvesting of photoexcitons. The performance of OPV cells based on BHJs depends critically on the morphological structure of the active layer blend on the nanoscale [224–226]. This explains the large amount of reports on this topic in the last few years, where morphological as well as further aspects of the BHJ blend were addressed. Investigations on the morphology were conducted by AFM [22], conductive AFM [227], scanning transmission electron microscopy (STEM) [228], TEM [122, 134, 229, 230] and TEM tomography [231]. Studies focusing on electric properties were performed with scanning tunneling spectroscopy [232] and SKPM [220, 233, 234], the crystalline structure was investigated using X-ray diffraction [136, 217]. Latest simulations on the BHJ blend morphology are presented in [235].

Especially in solution processed polymer OPV [236] there is a large set of parameter determining the crystallinity and morphology of the active layer blend [237, 238]. In vacuum processed small molecule solar cells BHJs are prepared by co-evaporation of donor and acceptor molecules [239]. As discussed in section 3.2.3, here the parameter space in terms of unintentional manipulations is much smaller than in solution processing. This in addition to the easy and manifold implementation of intentional manipulations such as vertical gradients of concentration [240–242] and crystallinity [243, 244] or thickness variations offers excellent conditions for both tailoring BHJ blends and addressing fundamental questions in OPV. The most prominent process parameter determining the morphological structure of co-evaporated BHJ blends are DA blend ratio, substrate and substrate temperature. Whereas the choice of DA blend ratio and substrate is often made under energetic considerations such as energy band alignment etc., the substrate temperature is the parameter most frequently varied to tailor the morphology of the blend. Although there is a broad consensus in the community that elevated substrate temperatures during co-evaporation induce significant changes in the blend morphology, the impact of these changes on PCE is still under debate. Whereas most of the studies report on a significant enhancement of PCE with substrate temperature [14–18], others find no change [19, 20] or even a decrease in PCE [21, 22]. Analytical tools were applied in these studies to reason the respective results. Still, a consistent theory providing a coherent explanation on this topic is still missing.

In this work we prepared $F_4ZnPc:C_{60}$ BHJ samples for a detailed investigation of the structure-function relationship in vacuum processed solar cell devices. $F_4ZnPc:C_{60}$ BHJ devices are known for exhibiting enhanced PCE due to substrate heating: Meiss et al. reported an increase in efficiency from 3.6% when processed at RT to 4.6% when processed at 104 °C. They investigated RT as well as heated blends with AFM and SEM and found significantly larger feature sizes in the topography of the heated samples. They concluded that this indicates an enhanced phase separation and generation of coherent percolation paths within the blend. XRD measurements on $F_4ZnPc:C_{60}$ blends did not show indications towards an enhanced crystallinity in one of the compounds. Thus they assigned the enhanced PCE to a more beneficial morphology of the blend neglecting the issue of blend crystallinity. Here, we used $F_4ZnPc:C_{60}$ because of the competitive performance and the known accessibility towards substrate heating as a model system to study the origin of PCE enhancement in heated small molecule organic solar cells. In a systematic study we characterized organic solar cells in different device architectures. Morphological as well as structural properties of the BHJ blend were investigated using energy-filtered TEM and TEM diffraction. One large benefit of this study is the close relation between both device and sample

5.2 C_{60} crystallinity dictates device efficiency in F_4ZnPc:C_{60} BHJ solar cells

preparation for analytics. Thus direct conclusions for the device physics can be drawn from the analytical findings.

5.2.2 Solar cell results on F_4ZnPc:C_{60} BHJ devices

In this section we present results F_4ZnPc:C_{60} OPV devices from conventional and inverted BHJ cells. For both the conventional and the inverted solar cells (almost) the same layer stack was used (but of course in inverse order). As discussed in section 2.2 the active layer morphology of these solar cells depends critically on the process conditions during co-evaporation. Here, we present a study on OPV cells with active layers processed on substrates at room temperature (RT) and 100 °C, which was found to yield the highest device performance for this material combination [17].

When speaking of solar cells processed at elevated substrate temperature we refer to the fabrication step of the active layer. Only after the deposition of the pristine F_4ZnPc (C_{60}) layer in the conventional (inverted) device architecture the substrate was heated to 100 °C. After the deposition of the BHJ blend and prior to the deposition of the pristine C_{60} (F_4ZnPc) layer in the conventional (inverted) device architecture, the heating was stopped and the sample cooled down about one and a half hour to a temperature < 60 °C.

5.2.2.1 Conventional BHJ devices

Figure 5.4 shows IV curves of conventional BHJ champion devices processed with the substrate at RT and 100 °C under AM1.5 illumination and in the dark. The performance of the cells is compared with data from Meiss et al. [17] in table 5.3. They optimized the active layer thicknesses as well as the doping ratios of the organic extraction layers. Before we discuss the effect of substrate heating on the performance, we compare our results on cells processed at RT with literature and data from bilayer systems. All devices presented were characterized in a glovebox under N_2 atmosphere.

We find that the absolute device performance achieved here is in the order of the reported values. Moreover, the superiority of the cells presented by Meiss et al. is exclusively based on a 17% higher V_{oc}. This is very likely due the tailored energy level alignment they achieved by using optimized doping levels within the organic transport layers. Our values for j_{sc} and FF do not

5 Structure-function relationship in F_4ZnPc/C_{60} solar cells

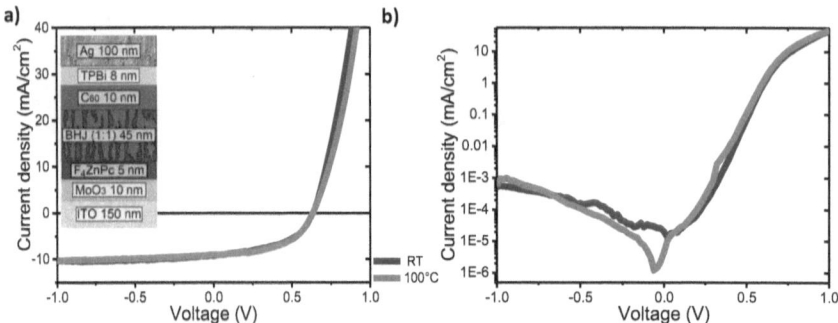

Figure 5.4: IV Characteristics of the $F_4ZnPc:C_{60}$ BHJ devices (best cells) in conventional device geometry processed with the substrate at RT (blue curve) and at 100 °C (red curve). a) IV characteristics under AM1.5 illumination. Inset: the used solar cell stack. b) IV characteristics under dark conditions.

exhibit significant deviations from literature, emphasizing the good extraction efficiency of the contacts used here. When comparing data presented here to data from (conventional) bilayers (see previous section), we find decreased values for V_{oc} and FF in the BHJ cells. The latter is likely caused by the hindered charge transport in the mixed BHJ phase: besides missing charge carrier percolation paths formed by single materials, TEM diffraction data shows that the growth of crystalline domains as present in the pristine bilayer films already at RT is suppressed in the BHJ (see section 5.2.3.2). The loss in V_{oc} is in line with reports on the poor electron blocking behaviour of the MoO_3 anode: because of the only 5 nm thin F_4ZnPc interlayer, more electrons than in the bilayer devices reach the hole extracting contact and recombine.

When comparing our dark IV curves to the ones reported from Meiss et al. we find a better diode behaviour for our cells. Whereas the dark current under forward bias is comparable ($j(+1\,\text{V}) = 42 \pm 5 \frac{\text{mA}}{\text{cm}^2}$ for our cells, $j_{\text{Lit}}(+1\,\text{V}) = 40 \pm 10 \frac{\text{mA}}{\text{cm}^2}$ for Meiss cells), under reverse bias the dark current is more than one magnitude lower for our cells ($j(-1\,\text{V}) = (5 \pm 2)10^{-4} \frac{\text{mA}}{\text{cm}^2}$ versus $j_{\text{Lit}}(-1\,\text{V}) = (2 \pm 1)10^{-2} \frac{\text{mA}}{\text{cm}^2}$ for Meiss cells). Thus the rectification efficiency of our solar cells is superior.

No significant effect of substrate heating on the device performance was found in the conventional BHJ architecture. This conclusion is based on a very solid statistical basis, as both the RT and the 100 °C samples could be

5.2 C_{60} crystallinity dictates device efficiency in $F_4ZnPc:C_{60}$ BHJ solar cells

	RT	100 °C	Meiss et al.
V_{oc} (mV)	628 ± 9	636 ± 9	730
j_{sc} (mA/cm^2)	9.1 ± 0.1	9.0 ± 0.1	9.2 ± 0.1
FF (%)	50.2 ± 2.2	48.1 ± 0.6	50 ± 1
Sat	1.14	1.13	1.07
η (%)	2.9 ± 0.1	2.8 ± 0.1	3.5 ± 0.1

Table 5.3: Characteristics of conventional $F_4ZnPc:C_{60}$ BHJ solar cells processed at RT and 100 °C substrate temperature. The values represent the average over the eight best cells. For comparison, literature values reported from Meiss et al. [17] for inverted OPV cells processed at RT are given. The temperature elevation during device fabrication does not lead to significant changes in the OPV performance. The saturation of the best cells shown in figure 5.4 is given: also here, no significant difference between RT and 100 °C samples is observed.

prepared with high (intra-batch) uniformity and (inter-batch) reproducibility. TEM studies discussed in section 5.2.3 reveal the device physics leading to the unexpected result of unchanged device performance under substrate heating.

5.2.2.2 Inverted BHJ devices

Figure 5.5 shows IV curves of inverted BHJ champion devices processed at RT and at 100 °C under AM1.5 illumination and in the dark. The characteristics of the cells are compared with data from Meiss et al. [17] in table 5.4. All devices were characterized in ambient using barrier foil for encapsulation.

For our inverted BHJ cells at RT we find good performance keeping up with our results on BHJ cells in the conventional architecture as well as with literature results. Comparing the characteristics of the inverted RT devices to our non-inverted ones we find no significant deviation. KP measurements demonstrate that the AlZnO treatment lowers the work function of the O_2 plasma treated ITO by about 1.3 eV to a value of 4.3 ± 0.1 eV, making the ITO/AlZnO:C_{60} interface a good electron extracting contact.

Unfortunately, our inverted BHJ cells processed at elevated substrate temperature could not keep up with literature results. Here the best device achieved is shown: the rectifying efficiency is significantly declined, leading to lower overall performance. Altogether, these cells exhibited huge variations in

5 Structure-function relationship in F_4ZnPc/C_{60} solar cells

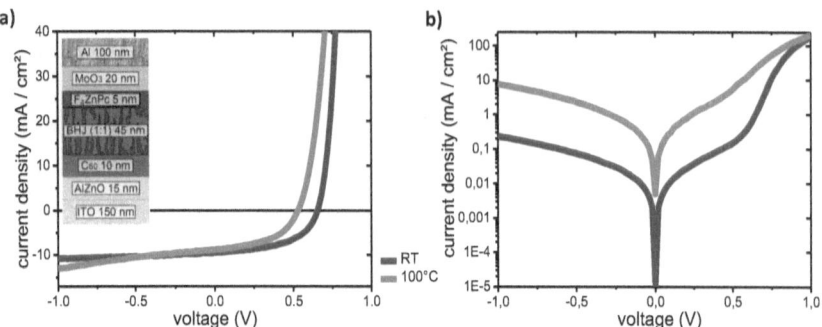

Figure 5.5: IV Characteristics of the $F_4ZnPc:C_{60}$ BHJ devices (best cells) in inverted device geometry processed with the substrate at RT (blue curve) and at 100 °C (red curve). a) IV characteristics under AM1.5 illumination. Inset: the used solar cell stack. b) IV characteristics under dark conditions.

performance (this both intra-/ and inter-batch-wise) with mostly very poor PCEs. Therefore a statistical treatment of the obtained results does not lead to meaningful conclusions here. As shown in table 5.4, Meiss et al. got an increase in efficiency of ≈ 17% when processing the inverted devices at elevated substrate temperature. We put large efforts in trying to reproduce these results. However, although several modifications on the stack were made we could not come up with similar results. Some of the modifications undertaken are discussed in the following.

To exclude negative effects of the annealing of the AlZnO layer in the UHV, we compared heated and non-heated inverted BHJ cells with non-plasma treated ITO as electron extracting contact. But also here substrate heating had a negative effect on device performance. We introduced a (p-conducting) TCTA buffer layer between MoO_3 and the active layer aiming to shield the latter from MoO_3 implantation, but the currents achieved were very low. Although TCTA looked promising in terms of its energy levels [245], the $F_4ZnPc/TCTA$ interface apparently hampers efficient hole transit. Also an increase of the pristine F_4ZnPc layer on top of the BHJ from 5 to 12 nm did not lead to elevated PCEs with substrate heating (but only to lower FFs).

Based on findings from TEM studies presented in section 5.2.3, we are convinced that the reported beneficial effect of substrate heating indeed exists for inverted BHJ devices. However, it seems that the device stacks used

5.2 C_{60} crystallinity dictates device efficiency in $F_4ZnPc{:}C_{60}$ BHJ solar cells

	RT	100 °C	RT_{Lit}	100 °C_{Lit}
V_{oc} (mV)	616 ± 25	580	730	680
j_{sc} (mA/cm^2)	9.2 ± 0.2	8.9	9.2 ± 0.1	10.0 ± 0.2
FF (%)	51 ± 1	51	50	60
Sat	1.14	1.49	1.07	1.10
η (%)	2.9 ± 0.2	2.7	3.5 ± 0.1	4.1 ± 0.1

Table 5.4: Characteristics of inverted $F_4ZnPc{:}C_{60}$ BHJ solar cells processed at different substrate temperatures and compared with literature values from Meiss et al. [17]. The values given at the first row represent the average of the eight best cells, in the second row only the best cell is shown. The temperature elevation during device fabrication changes OPV performance significantly. The enhanced saturation value for the heated sample documents on its poorer rectification.

here do not allow to produce reliable results. Most probably the evaporation of MoO_3 right onto the active layer causes this problem: it is known that MoO_3 penetrates deeply ($\gtrsim$ 20 nm) in underlying organic layers [246]. Also the heating induced change of the blend topography in combination with the strong MoO_3 penetration could account for the poor performance. AFM results presented in section 5.2.4 show that the blend roughness increases only slightly from 3.9 to 4.2 nm via substrate heating. However, the strongly differing topography could induce a modified growth of the pristine F_4ZnPc layer on top of the blend, probably less effective in shielding the blend from MoO_3 evaporation. Still, our data does not allow a final answer here.

5.2.2.3 Conclusion: solar cell results

We developed simple OPV stacks for efficient vacuum processed small molecule $F_4ZnPc{:}C_{60}$ solar cells in (conventional) bilayer as well as in BHJ (conventional and inverted) architecture.

The simple bilayer solar cell stack with MoO_3 coated ITO as hole extracting contact exhibited a PCE of 1.6±0.1%. This result keeps up with latest reports from literature [160].

Same applies to the BHJ solar cells we processed in conventional architecture: we achieved PCEs of about 2.9 ± 0.1% for devices prepared with the substrate kept at RT. Surprisingly, when preparing them at elevated substrate temperatures, the performance of these devices did not change significantly. This applies to all solar cell device parameter as well as to the diode characteristics extracted from IV curves in the dark. The results were highly reproducible.

For BHJ solar cells in inverted architecture we achieved good PCEs of about $2.9 \pm 0.2\%$ for devices prepared at RT. Unfortunately we could not reproduce reported results on inverted $F_4ZnPc:C_{60}$ BHJ cells processed at elevated substrate temperatures. Meiss et al. reported a significant increase in PCE for cells processed at 100 °C. Our cells exhibited poorer performance if the substrate was kept at this temperature during the deposition of the BHJ blend. Furthermore the obtained performances varied strongly, reproducibility was not given.

5.2.3 Revealing the BHJ nanostructure with analytical TEM

In this section we present results of a detailed transmission electron microscopy (TEM) study on $F_4ZnPc:C_{60}$ BHJ solar cells. In section 3.1.6 the fundamentals of TEM were discussed. We designed TEM samples mimicking the actual active layers of BHJ devices, so that phase separation and crystallinity of differently prepared BHJ solar cells could be studied via these TEM samples. The TEM data are correlated with the performance of BHJ solar cells. With this we could identify the C_{60} crystallinity as the main driving force of increased PCE in small molecule BHJ devices that are vacuum processed at elevated substrate temperature. However, the elevated temperatures alone are not sufficient for C_{60} crystallization. As present in inverted device architecture, an underlying pristine C_{60} layer providing crystallization seeds for the BHJ is mandatory for this process to start. Therefore the inverted device structure is preferential for the fabrication of high-performing C_{60}-based small molecule OPV. After discussing experimental details we present energy-filtered TEM studies on the phase separation of the BHJ blend. Afterwards the focus is shifted to the crystallinity of identical samples studied with TEM diffraction.

Because we did not succeed in the fabrication of inverted BHJ devices in a reproducible manner, we fall back on data reported by Meiss et al. [17]. Regarding the performance of conventional BHJ devices, we refer to our data. Both data sets are presented in the previous section 5.2.2.

5.2 C_{60} crystallinity dictates device efficiency in $F_4ZnPc:C_{60}$ BHJ solar cells

Here, a summary highlighting the major results of this study is given. The data presented were cross-checked with results from literature as well as from complementary methods and proved both highly reliable and coherent. These more detailed (and more critically) discussions on results, method and sample preparation can be found in the theses of Felix Schell [152] and Diana Nanova [37] as well as in our joint publication "Why inverted small molecule solar cells outperform their noninverted counterparts" [247].

Experimental details

All TEM samples were prepared by means of a floating process: organic $F_4ZnPc:C_{60}$ thin films were evaporated onto PEDOT:PSS coated ITO substrates. Exploiting the water solubility of PEDOT:PSS, the films deposited on top are delaminated using deionized water and picked up with a TEM grid which are mounted in the TEM for characterization. This preparation process implies that the sample undergoes water contact as well as exposure to ambient air. However, in a very detailed analysis Diana Nanova demonstrated that several results from literature on $F_4ZnPc:C_{60}$ samples not undergoing these contaminations could be reproduced with high accuracy. In the same manner she could exclude a significant impact of electron beam damage on the TEM results [37].

The organic layers were prepared as described in section 3.2.2. We used layer thicknesses (45 nm BHJ), deposition rates (10 Å/min) and material mixing ratios (1:1) identical to the ones used for solar cell devices. As for the devices, heating was applied when depositing the blend layer only. The pristine underlayers were deposited at RT. Thus highest relevance of the conclusions from TEM data for the devices is ensured. The stacks prepared for TEM studies are shown in figure 5.6. For noise reduction the C_{60} underlayer was reduced to 5 nm with respect to the inverted device, where we used 10 nm of C_{60}. The F_4ZnPc underlayer used here had a thickness of 5 nm, as in our conventional devices.

All TEM measurements were performed with a Carl Zeiss Libra 200 MC Cryo DMU.

5 Structure-function relationship in F_4ZnPc/C_{60} solar cells

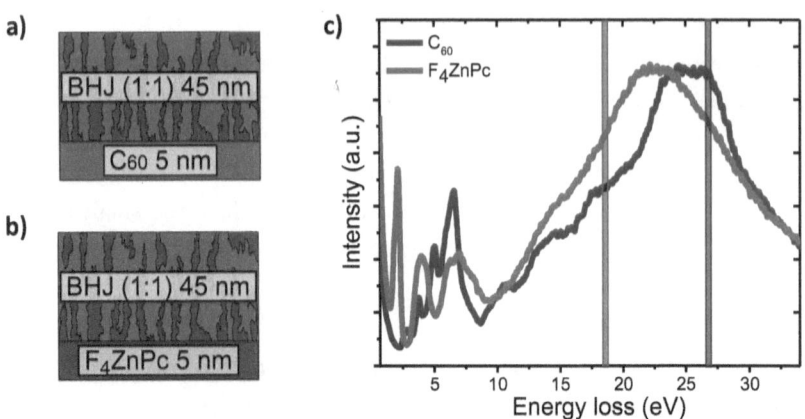

Figure 5.6: a), b) BHJ layer stacks used for TEM studies. a) TEM sample mimicking the inverted device structure. b) Mimic of the conventional structure. Samples were prepared both at RT and 100 °C. c) EEL spectra of pristine C_{60} (blue) and F_4ZnPc (red) thin films. The maxima of the plasmon peaks vary significantly, allowing meaningful plasmon peak mapping for this material combination. The loss energies with maximum contrast inversion are marked (18 eV for F_4ZnPc, 27 eV for C_{60}). The zero loss peak is blanked here.

5.2.3.1 Phase separation of $F_4ZnPc:C_{60}$ BHJs

To study the phase separation in $F_4ZnPc:C_{60}$ BHJ solar cells, we applied an analytical TEM series as described in 3.1.6: electron energy loss spectra (EELS) were taken to identify characteristic excitation energies for the two materials. A series of energy-filtered TEM (EFTEM) images was recorded. Plasmon peak mapping was used to extract two-class segmentations from these image series, providing a clear picture of the BHJ blend morphology.

Characteristic excitations of C_{60} and F_4ZnPc

To study characteristic excitations of the single materials, EELS for pristine C_{60} and F_4ZnPc were taken (see figure 5.6). These spectra show the intensity of the transmitted electron beam versus its energy (loss) in the low energy loss region. Here, electrons that underwent inelastic single-scattering-events are located, making this signal material-sensitive. The zero loss signal (originating from elastically scattered electrons) exceeds the peaks shown here by 2 or 3

5.2 C_{60} crystallinity dictates device efficiency in $F_4ZnPc:C_{60}$ BHJ solar cells

orders of magnitude in intensity and is excluded in this graph. In the loss region of $< 10\,\text{eV}$, mostly interband and HOMO-LUMO excitations of the respective materials are located. The broader excitations at higher energy losses stem from collective plasmon excitations of all valence electrons. These can be used to assign single pixels or regions in EFTEM images on BHJs to a certain material (or material class). In figure 5.6 the energies with maximum contrast inversion for the material system F_4ZnPc/C_{60} are marked by vertical bars. These two representative energies are used in the following to motivate the principles of electron spectroscopic imaging (ESI), where the whole series over all recorded loss energies is evaluated to achieve maximum material contrast. If studying energy-filtered TEM images at $18\,\text{eV}$ energy loss, F_4ZnPc-rich domains will appear bright because of the strong feature the material exhibits at this loss energy. C_{60}-rich domains on the other hand will appear dark. Same applies vice versa to energy-filtered TEM images at $27\,\text{eV}$. Thus it becomes possible to identify domains enriched with the different materials within the BHJ.

Real material contrast with energy-filtered TEM

In figure 5.7 EFTEM images of $18\,\text{eV}$, $27\,\text{eV}$ and zero-loss (ZL) taken from different BHJ samples are shown. The morphology of inverted (conventional) devices is addressed with BHJ layers grown on C_{60} (F_4ZnPc). EFTEM image series were taken for both configurations processed at RT and $100\,°\text{C}$. The EFTEM images from samples at RT show mostly noise at $18\,\text{eV}$ and $27\,\text{eV}$ loss energy, only in the ZL images certain features are visible. However, these could also stem from thickness variations and are therefore not further discussed here. Interesting is the formation of a distinctive morphological structure within the BHJ of both configurations if the substrate is heated to $100\,°\text{C}$. Already in the ZL images the emergence of a significant material agglomeration is visible. When analyzing the agglomerates spectroscopically at $18\,\text{eV}$ and $27\,\text{eV}$ loss energy, we find a clear contrast inversion for at least a significant part of them (some marked by red circles). This indicates that there is a significant phase separation between C_{60} and F_4ZnPc in the heated BHJ samples in both the conventional and inverted architecture. Thereby C_{60} agglomerates seem to drive the phase segregation: the domains of strong contrast inversion are found to appear dark at $18\,\text{eV}$ and bright at $27\,\text{eV}$.

These domains are surrounded by areas exhibiting reverse contrast *inversion*. However, these undergo a much lower shift in intensity. This indicates that there is a significant phase separation between C_{60} and F_4ZnPc in the heated BHJ samples in both the conventional and inverted architecture which is driven by strong C_{60} agglomeration.

Visualizing the BHJ phase separation

To investigate the morphology of the BHJs processed at 100 °C in more detail, a whole EFTEM series containing images from 2 to 35 eV (in 1 eV steps) is evaluated with plasmon peak mapping [229]. An Otsu threshold [248] ensuring maximal contrast between the plasmon peaks of the two materials is applied. Thus a two-class segmentation is achieved, revealing the phase separation in the BHJ. The results of the segmentation are shown in figure 5.8. There is a strong demixing in the BHJ processed at 100 °C for both cases. As already visible in the EFTEM images presented above, the BHJ exhibits stronger segregation when grown on F_4ZnPc. From power spectral density analysis features between 30 and 70 nm are present in the BHJ grown on F_4ZnPc, whereas the features in the BHJ on C_{60} are typically $\leq$ 30 nm. From analysis on the area ratios of F_4ZnPc and C_{60} in the segmentation arises that C_{60} is significantly underrepresented in both blends by $>$ 50%. This indicates that the F_4ZnPc rich domains still contain significant amounts of C_{60}.

Implications for solar cell performance

In EFTEM images and extracted segmentation graphs we found a strong effect of substrate heating to 100 °C onto the $F_4ZnPc:C_{60}$ BHJ morphology. Whereas both BHJ blends grown on C_{60} and on F_4ZnPc are not significantly phase segregated when processed at RT, at 100 °C they exhibit distinct phase segregation. This phase segregation seems to be driven by strong C_{60} agglomeration both for heated blends grown on C_{60} and F_4ZnPc. Thereby the conventional blend grown on F_4ZnPc exhibits bigger domains of 30 to 70 nm than the inverted blend grown on C_{60}, showing domain sizes of typically $\leq$ 30 nm. This backs assumptions from Meiss et al., who related increased PCEs in inverted $F_4ZnPc:C_{60}$ BHJ devices processed at 100 °C to a stronger phase separation, leading to well formed networks for both phases which offer

5.2 C_{60} crystallinity dictates device efficiency in $F_4ZnPc:C_{60}$ BHJ solar cells

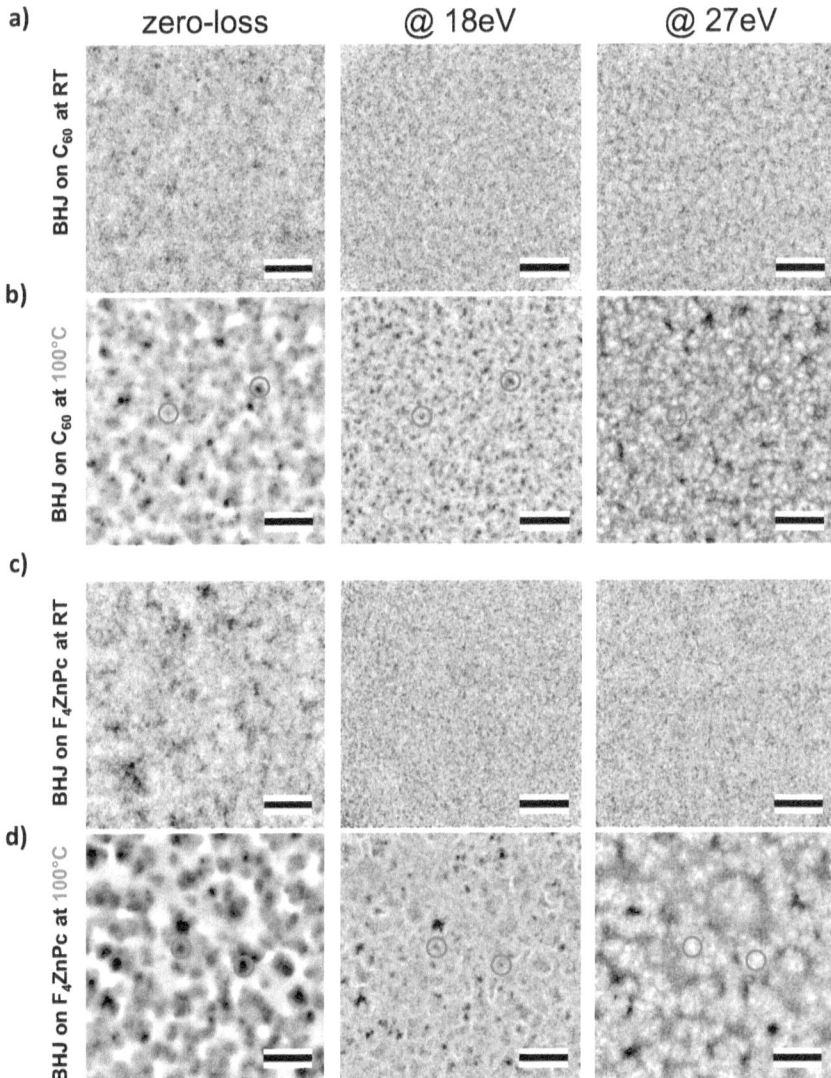

Figure 5.7: Energy-filtered TEM (EFTEM) images of BHJs deposited on different substrates and substrate temperatures. BHJs deposited on a), b) 5 nm C_{60}; c), d) 5 nm F_4ZnPc at RT and 100 °C. Shown are zero-loss as well as TEM images of 18 eV and 27 eV energy loss, characteristic for strong F_4ZnPc and C_{60} contrast respectively. RT samples exhibit a homogeneously intermixing, samples prepared at 100 °C exhibit strong agglomeration. The red circles mark areas of pronounced contrast inversion in the EFTEM images: they appear dark (bright) at 18 eV (27 eV) energy loss, indicating C_{60}-enriched areas. The scale bars correspond to 100 nm.

5 Structure-function relationship in F_4ZnPc/C_{60} solar cells

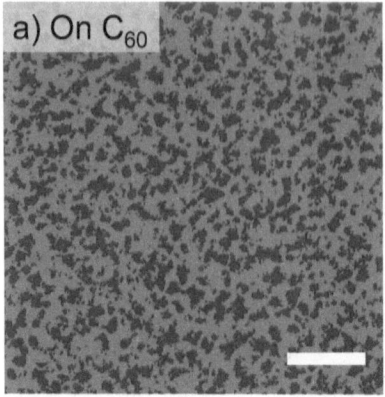
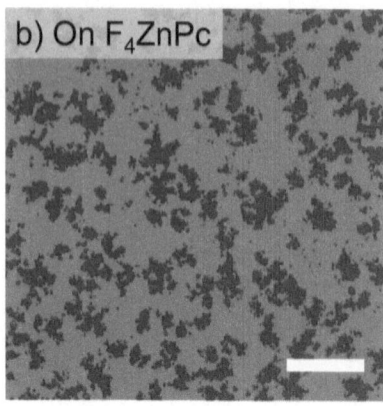

Figure 5.8: Two class segmentation of BHJs processed at 100 °C on a) C_{60} and b) F_4ZnPc. The plasmon peak mapping reveals agglomeration of F_4ZnPc (red) and C_{60} (blue). Features between 30 and 70 nm ($\leq$ 30 nm) are present in the BHJ grown on F_4ZnPc (C_{60}). The scale bars correspond to 100 nm.

coherent charge carrier percolation paths and suppress bimolecular recombination. However, this raises the question why we did not observe a similar improvement in our cells in conventional solar cell architecture: despite the strong morphological effect found here, no increase in solar cell performance is achieved for these cells.

5.2.3.2 Crystallinity in F_4ZnPc:C_{60} BHJs

In the previous section we learned that the morphology of the F_4ZnPc:C_{60} blend layer is apparently not the only parameter determining the device performance. To address this issue, we performed TEM diffraction on the identical samples that were used for the morphological studies. For details on the method see 3.1.6.

Strong templating for heated blends on C_{60}

The patterns of TEM diffraction on BHJ blends grown on the different substrates and temperatures are shown in figure 5.9 a). All patterns are characterized by three distinct rings. For both blends processed at RT these patterns appear rather similar to each other and overall less accentuated

5.2 C_{60} crystallinity dictates device efficiency in $F_4ZnPc:C_{60}$ BHJ solar cells

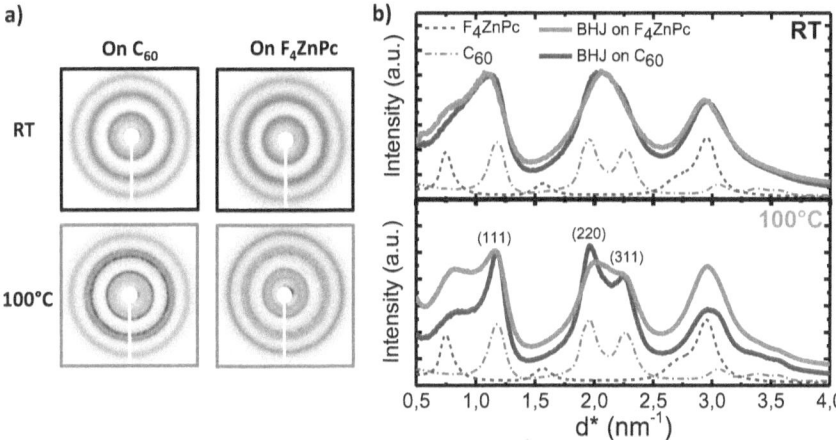

Figure 5.9: TEM diffraction data of BHJ blends grown under different conditions. a) TEM diffraction patterns of $F_4ZnPc : C_{60}$ blends grown on F_4ZnPc and C_{60} at both RT and 100 °C. b) Radial profiles extracted from the presented patterns and from pristine layers processed at RT. The diffraction profiles of blends grown at RT are very similar for both cases. At 100 °C the C_{60} crystallinity of the blend grown on C_{60} increases significantly, visible in the formation of the distinct C_{60} peaks (111), (220) and (311). For the blend on F_4ZnPc only a minor crystallization effect is observed.

when compared to the 100 °C samples. The diffraction patterns from the heated samples indicate that the blend crystallinity depends critically on the underlying material: whereas the main change of the blend on F_4ZnPc is the emergence of speckles in the regions of the rings, in the blend on C_{60} there is a strong signal increase on the edges of these rings. The trends described become more comprehensible in figure 5.9 b), where radial profiles of the patterns are shown aside of profiles from pristine F_4ZnPc and C_{60} films processed at RT. We find that the profiles of the RT blends are almost identical. The profiles from the blends processed at 100 °C indicate the strong impact of the underlying substrate: whereas the blend on F_4ZnPc exhibits only slight indications of enhanced F_4ZnPc crystallization (for example at $0.45\,\mathrm{nm}^{-1}$), the C_{60} fraction of the blend grown on C_{60} is of high crystalline order. All peaks that appear in the pristine C_{60} layer are present and well pronounced in the heated blend on C_{60} as well. Comparing the results from both heated blends it becomes apparent that there are *two* prerequisites that have to be fulfilled if an improvement of the BHJ blend quality is desired: besides the supply of sufficient thermal energy via heating, there is need of a templating underlayer which induces crystalline growth.

Implications for solar cell performance

Last section closed with the question why the formation of distinctive phase separation of both conventional and inverted F_4ZnPc:C_{60} BHJ solar cells has a different impact on device performance: whereas it is beneficial in the inverted device structure, performance is not increased in conventional devices. Here we found that there is a significantly enhanced C_{60} crystallinity present in the BHJ blend on C_{60}, mimicking the architecture of inverted cells. Thus it becomes clear that the increase in C_{60} crystallinity is accountable for the enhanced performance in the heated inverted devices. The temperature induced increase in crystallinity is much smaller for the blends grown on F_4ZnPc, mimicking conventional devices. This leads us to the conclusion that the crystallinity within the BHJ blend plays a way more prominent role for device efficiency than expected.

5.2 C_{60} crystallinity dictates device efficiency in $F_4ZnPc:C_{60}$ BHJ solar cells

5.2.4 Prospect on the 3rd dimension-AFM studies on BHJ blends

In TEM the vertical dimension of the sample is projected into the horizontal plane. Here we access this dimension (at least at the blend surface) with AFM. We use the topographical information to cross-check TEM results and broaden our view on the BHJ morphology. For this, $F_4ZnPc:C_{60}$ BHJ blends prepared analog to the TEM samples are characterized with AFM. Morphological changes due to heating and typical topographic feature sizes are discussed in the light of the TEM findings.

The author thanks Christian Müller for support with the AFM measurements at Kirchhoff-Institut für Physik/Heidelberg.

Experimental details

As substrate we used plasma-cleaned ITO. BHJ blends were deposited as described in the previous section: pristine thin films of 5 nm F_4ZnPc (C_{60}) were followed by a 45 nm nm BHJ blend for mimicking conventional (inverted) devices. The stacks were illustrated in figure 5.6 of the previous section. Deposition rates (10 Å/min) and material mixing ratios (1:1) were identical to the ones used for TEM samples and solar cell devices. The pristine layers for the conventional and the inverted samples respectively were applied simultanously at RT. Also the blend layers for the RT and the 100 °C samples were deposited in the same process. All samples were characterized in the same session and using the same cantilever. Thus a high degree of comparability between the samples is ensured. The samples were characterized at ambient air with a MultiMode AFM from Bruker using tapping mode AFM. AFM image analysis was performed with the free and open source software *Gwyddion* [249].

Experimental results

In figure 5.10 AFM micrographs of the differently processed $F_4ZnPc:C_{60}$ blends are shown. BHJ samples with a pristine F_4ZnPc (C_{60}) underlayer mimicking conventional (inverted) devices processed at RT and 100 °C are compared. In terms of layer composition and process conditions, they are identical to the samples studied with TEM presented in figure 5.7.

5 Structure-function relationship in F_4ZnPc/C_{60} solar cells

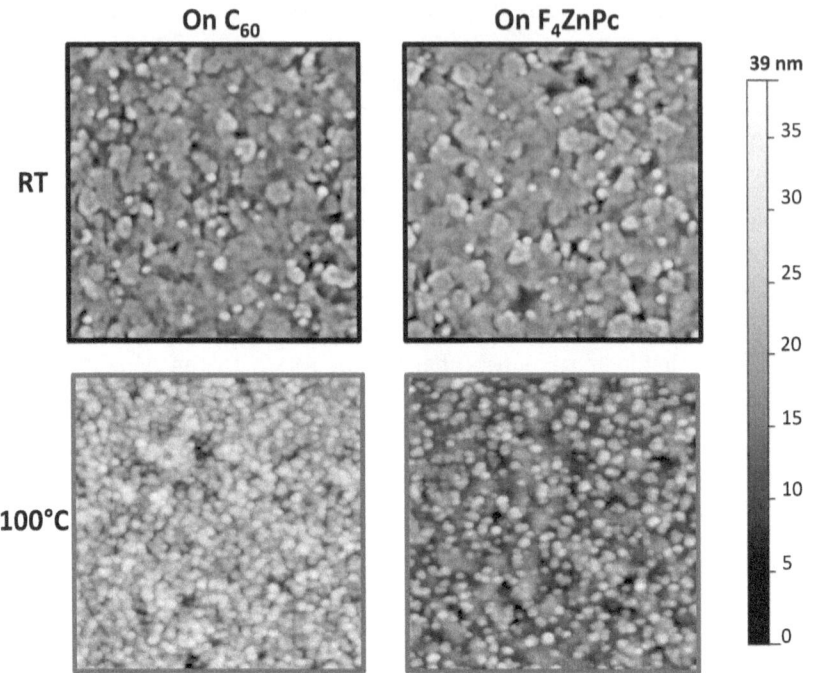

Figure 5.10: AFM micrographs ($2 \cdot 2\,\mu m^2$) from $F_4ZnPc:C_{60}$ BHJ blends grown under different conditions. Left: blend grown on C_{60}; right: blend grown on F_4ZnPc. Top (bottom) row: RT (100 °C) samples. The z-scale of all images is 39 nm.

When comparing the AFM images of the conventional and inverted samples prepared at RT (first row), we find that they are very similar to each other both qualitatively and quantitatively. For both we find a root mean square (rms) roughness of 3.9 nm and similar topography. The partly flake-like structure recalls the nanostructure of the ITO substrate as found in 5.2 buried under the organics. Besides this, some spheres are visible with a lateral extension of some tens of nm and heights of < 15 nm. These spheres seem not to have a counterpart in the ZL EFTEM images shown in figure 5.7. The blends processed at 100 °C vary significantly. We find that the conventional blend (grown on F_4ZnPc) exhibits a pronounced formation of spheres with a lateral extension of some tens of nm and heights of < 20 nm emerging from a rather smooth background. The inverted blend (grown on C_{60}) seems to be built

5.2 C_{60} crystallinity dictates device efficiency in F_4ZnPc:C_{60} BHJ solar cells

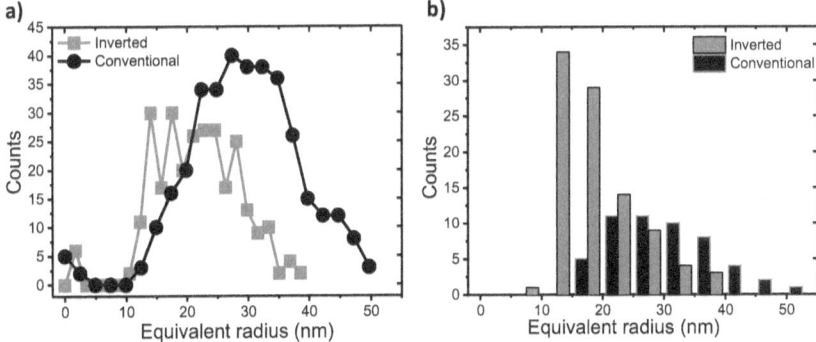

Figure 5.11: Equivalent radii of grains from the AFM micrographs of the heated blends presented in figure 5.10. a) Automized evaluation based on watershed algorithm. b) Manual evaluation. The blend in conventional structure (on F_4ZnPc) exhibits bigger grains than the blend in inverted structure (on C_{60}). From the evaluation of both plots we find average radii of a) 22 ± 9 nm (31 ± 8 nm) and b) 18 ± 7 nm (29 ± 8 nm) for the inverted (conventional) blend. In (a) more grains are detected in the conventional sample because the algorithm applies better to the more distinct grains present in this sample.

up by a multitude of smaller, partly coalesced spheres. The rms roughness is 4.5 nm (4.2 nm) for the conventional (inverted) blend. These observations are in line with the results from TEM measurements: especially in the case of the heated conventional blend the spheres visible here correspond to the extended areas exhibiting higher gray values in the ZL EFTEM images on the analogous sample. Also the smaller feature sizes observed for the inverted 100 °C blend fits the results of the TEM study.

We further analyzed the AFM images of both samples processed at elevated substrate temperature to quantify these features. We marked grains in the AFM images with (i) a watershed algorithm implemented in Gwyddion and (ii) manually to extract the equivalent radii of the features [2]. In the manual evaluation areas of about 20% of both images shown in 5.10 were considered. As expected from the AFM micrographs, more grains per area were found for the blend in inverted architecture: in plot (b) 94 (64) grains were considered for the inverted (conventional) sample. In the automized evaluation applied

---

[2] An additional algorithm based on thresholds of height, slope and curvature of features in the AFM image was tested, but watershed achieved much better results.

on the entire micrographs more grains were found for the conventional sample, because the watershed algorithm is more effective in the identification of its more distinct grains. However, the algorithm provides a sufficient number of grains for both samples: in plot (a) 278 (352) grains were considered for the inverted (conventional) sample. Errors are given by the standard deviation. The typical feature sizes found in the manual evaluation were slightly smaller, probably because here less coalesced grains were taken into account. Overall the automized and manual evaluation are in very good agreement.

The results are presented in figure 5.11: from the automized (manual) evaluation we find average radii of $22 \pm 9\,\mathrm{nm}$ ($18 \pm 7\,\mathrm{nm}$) for the blend in inverted architecture and $31 \pm 8\,\mathrm{nm}$ ($29 \pm 8\,\mathrm{nm}$) for the blend in conventional architecture. Because of their good agreement, for further discussion we use average radii of 20 ± 8 and $30 \pm 8\,\mathrm{nm}$ for the inverted and conventional blend, which result in typical feature sizes of 40 ± 16 and $60 \pm 16\,\mathrm{nm}$ in diameter respectively. We find that this analysis yielding significant larger features for the blend in conventional architecture is in line with the qualitative observations discussed above.

5.2.4.1 Conclusion: AFM studies on BHJ blends

The blends are very similar when grown at RT. Blends deposited at $100\,°C$ on F_4ZnPc exhibit a higher roughness and stronger feature formation than blends grown on C_{60}. From the two-class segmentation of EFTEM measurements feature sizes between 30 and 70 nm were determined for the blend grown on F_4ZnPc at $100\,°C$. The BHJ on C_{60} exhibited feature sizes of typically $\leq 30\,\mathrm{nm}$ (see 5.2.3.1). With AFM we got typical feature sizes of $60 \pm 16\,\mathrm{nm}$ and $40 \pm 16\,\mathrm{nm}$ for the heated blends on F_4ZnPc and C_{60}. The slightly higher values determined here from AFM micrographs could be caused by the resolution limiting effect of the finite cantilever tip size: because of the feature's convolution with the tip, they appear bigger in the AFM image. In TEM there is no equivalent limiting effect on the given scale. Overall, there is a good agreement between the feature sizes determined by TEM and AFM. This should not be taken for granted, as the information used stem from different sample properties: spectroscopic information from the entire bulk in EFTEM on the one and topographic information from the surface in AFM on the other hand. This raises the question whether the topography signal enclosed in the TEM images of the ESI series influences the two-class segmentation. Or the other way around: whether the phase segregation observed there dictates the topography of the BHJ surface determined by AFM. Here we think the latter is the case: the contrast inversion present in the ESI series makes a strong point backing the segmentation results obtained with EFTEM. This being said, it seems that no significant correction of the picture obtained by analytical TEM has to be made, emphasizing the potency of EFTEM characterization. Still, this also implies that AFM allows for an easy and fast screening of the phase separation in vacuum processed BHJ blends: enhanced phase separation alters the topography significantly. However, further studies containing TEM diffraction are inevitable to address all parameter determining the quality of BHJ active layers.

5.2.5 Discussion: Structure-function relationship

The aim of this study was the profound understanding of the structure-function relationship in BHJ small molecule organic solar cells. The well-known material system $F_4ZnPc:C_{60}$ was used as model system to investigate the interplay between device performance and both morphology and crystallinity of the active layer blend. BHJ blend deposition at elevated substrate temperatures is a method widely used to optimize device performance. This however with mixed results, because a universal model providing explicit device design criteria was still missing. We installed a substrate heating unit in the organic deposition chamber and prepared BHJ solar cell devices as well as freestanding BHJ blends mimicking the active layers of devices for TEM studies under defined processing conditions. Energy-filtered TEM and TEM diffraction was applied to reveal morphology and crystallinity of the differently prepared blends. Based on the results of this study we can draw a clear picture of the mechanism underlying the PCE enhancement in small molecule OPV due to substrate heating and supply defined design criteria for high-performing OPV. Our main findings are depicted in figure 5.12. We find homogeneously intermixed and amorphous blends of F_4ZnPc and C_{60} when depositing the BHJ at RT. If the substrate is heated to 100°C during the deposition process of the blend, we observe strong phase segregation irrespective of the substrate. However, enhanced crystallinity is found in the inverted device structure only: the pristine C_{60} underlayer present in these devices is mandatory to

5 Structure-function relationship in F_4ZnPc/C_{60} solar cells

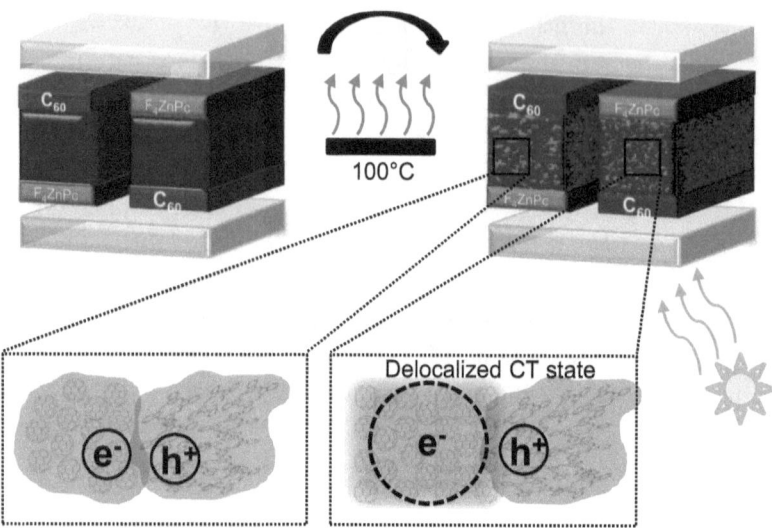

Figure 5.12: Graphical abstract summarizing the main findings of this chapter: opposite to blends deposited at RT (left hand side), the deposition of blends at elevated substrate temperature induces strong phase separation between F_4ZnPc and C_{60}. However, only if the blend is deposited onto a C_{60} underlayer this leads to enhanced crystallinity giving rise to higher delocalization of the CT electron. Figure reprinted from [37].

initiate the process of C_{60} crystallization. This C_{60} crystallinity is the main factor pushing the performance of the heated devices. Phase separation is necessary as well and driven by C_{60} agglomeration. Of course, the formation of sufficiently large C_{60} domains is a prerequisite for the crystallization. In the conventional device structure, there is no analogue to this templating effect: the pristine F_4ZnPc underlayer does not give rise to significantly stronger F_4ZnPc crystallization in the blend.

Before discussing these conclusions in the context of latest findings in literature, we want to motivate our reasoning in the light of the device data presented here. Recalling the enormous transformation of the conventional blend morphology due to substrate heating from homogeneous intermixing to distinct phase separation it is very surprising that there is no impact on

5.2 C_{60} crystallinity dictates device efficiency in $F_4ZnPc:C_{60}$ BHJ solar cells

device performance at all. The formation of the phase separation yields an interconnected network of the respective phases. If the homogeneously intermixed nature of the BHJ deposited at RT indeed constitutes the bottleneck of these devices, they should exhibit significantly enhanced performance when deposited at 100 °C: the interconnected network in the blend should lower non-geminate recombination (enhancing j_{sc}) and the voltage-dependence of charge carrier extraction (enhancing FF). However, we did not find any indication for this in the performance of conventional BHJ devices. On the other hand, we also rule out here that the indeed positive effect of the phase segregated morphology is counterbalanced by other effects. For example, a higher number of shortcuts or enhanced hole recombination at the top contact could be caused by the higher roughness. However, the former should lead to a inferior diode efficiency visible in the dark IV curve, the latter should lead to a lower V_{oc}. We did not find any indication for both of them. In conclusion, we think that the independence of the device performance on the phase separation in the blend indicates that inefficient charge extraction because of non-geminate recombination and low mobility does *not* constitute the bottleneck towards improved device efficiency in the conventional device structure, but rather the poor charge separation efficiency at the DA interface. This is backed by recent publications reporting on efficient charge extraction and exciton quenching, but still low internal quantum efficiencies found for different state-of-the-art OPV systems, indicating that strong geminate recombination at the DA interface is the limiting factor in these solar cells [40, 250].

This being said, we now focus on the main difference found between the device architectures: the significant increase in C_{60} crystallinity. Very recently, Devizis et al. found in ultrafast electroabsorption spectroscopy measurements on small molecule organic solar cells that "interfacial CT states split into free charges mainly by electron escape from the coulomb potential. The motion of holes in the small molecule donor material during the charge separation time is found to be insignificant." [38]. This challenges previous views overestimating the contribution of donor crystallinity [251] but is in line with both our findings and that of several other groups published in the last two years which identified crystalline fullerene domains offering delocalized states for CT electrons as the main driving force for highly efficient OPV [39–41]. Bartelt et al. reported similar findings on $ZnPc:C_{60}$ blends already in 2011 [252]. We want to mention here that the cited reports cover a variety of different OPV material combinations that are both solution [39] and vacuum [38, 41, 252] processed. This indicates that the underlying principle is of fundamental importance for efficient OPV.

These highly coherent results from literature strengthen our findings that C_{60} crystallinity is the crucial parameter on the road to efficient vacuum processed small molecule solar cells. For distinct C_{60} crystallization in these cells both sufficient thermal energy enhancing diffusivity and a pristine C_{60} underlayer initiating the crystallization process have to be provided (see figure 5.12). The advantage of our over the other studies is the close relation of fundamental insight and device fabrication, allowing for the derivation of design criteria for highly efficient vacuum processed OPV. These device design criteria are:

- Inverted device architecture is the prerequisite for highest performing devices.

- The "electron transport section" (often containing more than one homogeneous layer) has to be terminated by a layer of pristine C_{60}, providing crystallization seeds. This is emphasized here, because in many high performing OPV stacks doping molecules are co-evaporated with C_{60} to increase the overall mobility of the electron transport layer [253–255]. The impact of doping molecules on C_{60} crystallinity should be critically checked before employing the doped layer into the stack.

- Sufficient thermal energy has to be provided for optimum blend growth. Which temperature is *sufficient* has to be investigated for every particular material combination. However, for most material combinations the optimum temperature is in the range of $100 \pm 20\,°C$. Further studies on this topic were performed at iL by Christian Willig in the scope of his bachelor thesis [256]. He used the merocyanine HB194 [257] characterized by a higher dipole moment than F_4ZnPc. It was assumed that this dipole moment would induce sufficient phase separation already at RT. However, this was not confirmed in his studies. As recently found, the merocyanines "tend to pack tightly in antiparallel dimers" [258], summing up to a neglecting dipole moment. Thus the absence of the expected phase separation in the BHJ blend becomes reasonable.

6 Conclusion and outlook

In this thesis two main topics were investigated by means of characterization techniques with high lateral resolution. SKPM and analytical TEM were utilized to reveal electronic and structural features on the nanoscale which were correlated to the macroscopic device performance of different types of F_4ZnPc/C_{60} small molecule organic solar cells.

In the first part of this thesis the electric potential distribution of operating F_4ZnPc/C_{60} bilayer solar cells was studied. For this, solar cells with suitable layer thicknesses and varied hole extracting contacts were characterized with cross-sectional SKPM. In preliminary layer-by-layer studies with in-situ KP and PES, energy band alignment and chemical as well as electronic properties of the different hole extracting contacts were examined. We found a beneficial band bending for hole (electron) extraction at the F_4ZnPc/TCO ($C_{60}/TPBi/Ag$) interface and a disadvantageous band alignment at the F_4ZnPc/C_{60} DA interface. XPS studies revealed that the O_2 plasma treatment of the ITO substrate, which accounts for the good electronic interface alignment with the high-lying donor F_4ZnPc, activates the ITO surface and cancels its degeneration. F_4ZnPc molecules impinging on the activated surface are not (only) physisorbed, but interact chemically with the surface and decompose. This is not the case if good electronic interface alignment is achieved by coating the ITO with MoO_3, where no indications on chemical interactions of F_4ZnPc and substrate were found. In conclusion, both the O_2 plasma treated ITO and the MoO_3 coated ITO hole extracting contact exhibited good energy band alignment. However, an adverse decomposition of the first F_4ZnPc monolayer was found for the plasma treated ITO contact.

To examine the effect of FIB exposure on cross-sectional SKPM studies we correlated simulation and SKPM data from solar cell cross sections prepared using the different FIB ions He, Ne and Ga. There are large inhomogeneities

6 Conclusion and outlook

regarding the implantation of FIB ions in the solar cell cross sections. However, in the SKPM data no indications towards a systematic dependence between the electric potential distribution of the cell and these inhomogeneities were found. In particular, the He and Ne as well as the Ga FIB preparation did not lead to electric potential distributions under short circuit conditions that match the expectations from preliminary KP and PES studies.

F_4ZnPc/C_{60} bilayer solar cells with varied hole extracting contact were studied with in-operando SKPM under illumination and applied bias voltages. Although the energy band alignments were found similar under equilibrium conditions in UPS studies, the in-operando electric potential distributions of solar cells with O_2 plasma treated and MoO_3 coated ITO contacts varied strongly. At the maximum power point, the solar cell with MoO_3 coated ITO contact exhibited bulk limited operation with no potential loss in the contact region. The solar cell with O_2 plasma treated ITO contact exhibited a distinct contact limitation, which can be attributed to the decomposed F_4ZnPc donor layer right at the contact interface. In studies under applied bias voltages it was found that the transport barrier at the plasma treated ITO/F_4ZnPc interface can be described by a constant contact resistance $R_{contact}$, at least for the voltage range probed here ($V_{applied} \leq \pm 2V$). The ohmic behavior of the MoO_3/F_4ZnPc (and C_{60}/TPBi/Ag) interface was exploited to study the bulk transport properties of F_4ZnPc and C_{60}. It was found that the electron mobility of C_{60} is about 2 orders of magnitude higher than the hole mobility of F_4ZnPc. In conclusion, these results prove that the electric potential distribution of operating organic electronic devices can hardly be predicted from equilibrium analysis such as layer-by-layer studies with in-situ UPS. Furthermore it is demonstrated that in-operando SKPM studies are very sensitive towards the contact properties of organic electronic devices, making it a powerful screening tool in device analysis.

In the second part of this thesis the interplay of solar cell performance and active layer microstructure was studied.

In-situ AFM and XPS was used to study the growth of F_4ZnPc/C_{60} bilayer solar cells. Stranski-Krastanow growth was found for both the MoO_3/F_4ZnPc and the F_4ZnPc/C_{60} interface: after the formation of closed layers small islands of some ten nm are formed. Furthermore was shown that the TPBi capping layer reduces the surface roughness of the active layer by more than

50%. This smoothening ensures minimal interface area between active layer and Ag top cathode and hence minimizes recombination losses at the metal contact.

In a further study the impact of substrate heating during co-evaporation on the active layer morphology in $F_4ZnPc{:}C_{60}$ bulk heterojunction (BHJ) solar cells was examined. For this, active layer blends mimicking solar cell devices were investigated with analytical TEM. The results on material distribution and crystallinity were correlated with solar cell device data. For the TEM studies, BHJ blends were processed on pristine F_4ZnPc and C_{60} layers to mimic non-inverted (conventional) and inverted device architectures. The morphology of the blends in both configurations changes from homogeneously intermixed to phase segregated on a scale of some ten nm when the BHJ is processed at elevated (100 °C) substrate temperature. However, only in the inverted devices this lead to an improved power conversion efficiency, no improvement was found for the conventional devices. By means of TEM diffraction it was demonstrated that the fullerene agglomerates exhibit enhanced ordering when the blend is processed at elevated substrate temperature on a C_{60} underlayer, as present in inverted devices. If a F_4ZnPc underlayer is used, the agglomerates in the blend remain widely amorphous. We could demonstrate that the substrate-induced fullerene ordering is responsible for the enhanced power conversion efficiencies of inverted small molecule organic solar cells processed at elevated temperatures. The pure and ordered C_{60} domains lead to stronger delocalization of the charge transfer state, improving free charge carrier generation and performance. This finding explains why (i) record efficiencies of 8.3% in single junction cells were reported in inverted device architecture only [255] and (ii) literature reports on this topic came to mixed results. The effect of the underlayer as present in actual devices was not considered in the discussion on the BHJ blend morphology so far. In conclusion, we demonstrated that C_{60} crystallinity is the crucial parameter for efficient small molecule organic solar cells and that its formation necessitates the supply of both sufficient thermal energy and a crystallization-inducing substrate.

These results on F_4ZnPc/C_{60} solar cells raise the question whether there are material systems that exhibit enhanced crystallinity in both the acceptor *and* the donor material. To achieve crystallization in both phases during co-evaporation of the blend there is need of materials exhibiting similar behavior regarding diffusion and crystal formation in both phases at a specific substrate temperature.

A promising approach for increasing the power conversion efficiency in organic solar cells is the replacement of fullerenes. Fullerenes have non-ideal optical band gaps with respect to the solar spectrum, leading to low absorption and photocurrents. Furthermore they give rise to high recombination losses at the DA interface, thus limiting V_{oc} [218, 259]. Novel subphthalocyanines used as donor molecules allow high V_{oc} and power conversion efficiencies of 6.9% even in planar heterojunction architectures [260]. The replacement of fullerenes is also intended because they hamper a defined molecular packing at the donor acceptor interface, which is found crucial for device performance [154, 218, 261, 262]. Cnops et al. presented a fullerene-free small molecule organic solar cell which employed a cascade-energy-level alignment leading to power conversion efficiencies of 8.4% in a three-layer device architecture [46]. Probably solar cells exploiting this principle could reach significantly higher efficiencies if they would be realized using materials of higher crystalline order.

These results underline the large potential of organic photovoltaics and demonstrate that power conversion efficiencies competitive with established technologies are in reach.

Bibliography

[1] J. Koenigsberger and K. Schilling. Über Elektrizitätsleitung in festen Elementen und Verbindungen. I. Minima des Widerstandes, Prüfung auf Elektronenleitung, Anwendung der Dissoziationsformeln. *Annalen der Physik*, 337(6):179–230, January 1910.

[2] Max Volmer. Die verschiedenen lichtelektrischen Erscheinungen am Anthracen, ihre Beziehungen zueinander, zur Fluoreszenz und Dianthracenbildung. *Annalen der Physik*, 345(4):775–796, January 1913.

[3] W. Helfrich and W. G. Schneider. Recombination Radiation in Anthracene Crystals. *Physical Review Letters*, 14(7):229–231, February 1965.

[4] M. Pope, H. P. Kallmann, and P. Magnante. Electroluminescence in Organic Crystals. *The Journal of Chemical Physics*, 38(8):2042–2043, April 1963.

[5] D. F. Williams and M. Schadt. A simple organic electroluminescent diode. *Proceedings of the IEEE*, 58(3):476–476, March 1970.

[6] C. K. Chiang, C. R. Fincher, Y. W. Park, A. J. Heeger, H. Shirakawa, E. J. Louis, S. C. Gau, and Alan G. MacDiarmid. Electrical Conductivity in Doped Polyacetylene. *Physical Review Letters*, 39(17):1098–1101, October 1977.

[7] The Nobel prize in chemistry 2000. http://www.nobelprize.org/nobel_prizes/chemistry/laureates/2000/, April 2016.

[8] C. W. Tang. Two-layer organic photovoltaic cell. *Applied Physics Letters*, 48(2):183–185, January 1986.

[9] C. W. Tang and S. A. VanSlyke. Organic electroluminescent diodes. *Applied Physics Letters*, 51(12):913–915, September 1987.

[10] A. Tsumura, H. Koezuka, and T. Ando. Macromolecular electronic device: Field effect transistor with a polythiophene thin film. *Applied Physics Letters*, 49(18):1210–1212, November 1986.

[11] Data sheet Heliafilm from Heliatek. http://www.heliatek.com/en/heliafilm/technical-data, April 2016.

[12] Rebecca Saive, Michael Scherer, Christian Mueller, Dominik Daume, Janusz Schinke, Michael Kroeger, and Wolfgang Kowalsky. Imaging the Electric Potential within Organic Solar Cells. *Advanced Functional Materials*, 23(47):5854–5860, December 2013.

[13] Rebecca Saive, Christian Mueller, Janusz Schinke, Robert Lovrincic, and Wolfgang Kowalsky. Understanding S-shaped current-voltage characteristics of organic solar cells: Direct measurement of potential distributions by scanning Kelvin probe. *Applied Physics Letters*, 103(24):243303, December 2013.

[14] Fan Yang, Max Shtein, and Stephen R. Forrest. Controlled growth of a molecular bulk heterojunction photovoltaic cell. *Nature Materials*, 4(1):37–41, January 2005.

[15] Steffen Pfuetzner, Jan Meiss, Annette Petrich, Moritz Riede, and Karl Leo. Thick C60:ZnPc bulk heterojunction solar cells with improved performance by film deposition on heated substrates. *Applied Physics Letters*, 94(25):253303, June 2009.

[16] Christoph Schünemann, David Wynands, Lutz Wilde, Moritz Philipp Hein, Steffen Pfützner, Chris Elschner, Klaus-Jochen Eichhorn, Karl Leo, and Moritz Riede. Phase separation analysis of bulk heterojunctions in small-molecule organic solar cells using zinc-phthalocyanine and c60. *Physical Review B*, 85(24):245314, June 2012.

[17] Jan Meiss, Andre Merten, Moritz Hein, Christoph Schuenemann, Stefan Schäfer, Max Tietze, Christian Uhrich, Martin Pfeiffer, Karl Leo, and Moritz Riede. Fluorinated Zinc Phthalocyanine as Donor for Efficient

Vacuum-Deposited Organic Solar Cells. *Advanced Functional Materials*, 22(2):405–414, January 2012.

[18] Bregt Verreet, Paul Heremans, Andre Stesmans, and Barry P. Rand. Microcrystalline Organic Thin-Film Solar Cells. *Advanced Materials*, 25(38):5504–5507, October 2013.

[19] Peter Peumans, Soichi Uchida, and Stephen R. Forrest. Efficient bulk heterojunction photovoltaic cells using small-molecular-weight organic thin films. *Nature*, 425(6954):158–162, September 2003.

[20] S. Ouro Djobo, L. Cattin, M. Morsli, A. Godoy, F. R. Diaz, M. A. del Valle, and J. C. Bernede. Effect of the Substrate Temperature on the Performance of Small Molecule Organic Solar Cells. volume 1391, pages 251–253. AIP Publishing, 2011.

[21] Konstantinos Fostiropoulos and Wolfram Schindler. Donor-acceptor nanocomposite structures for organic photovoltaic applications. *physica status solidi (b)*, 246(11-12):2840–2843, December 2009.

[22] Christian Koerner, Chris Elschner, Nichole Cates Miller, Roland Fitzner, Franz Selzer, Egon Reinold, Peter Bäuerle, Michael F. Toney, Michael D. McGehee, Karl Leo, and Moritz Riede. Probing the effect of substrate heating during deposition of DCV4t:C60 blend layers for organic solar cells. *Organic Electronics*, 13(4):623–631, April 2012.

[23] Markus Schwörer and Hans Christoph Wolf. *Organic Molecular Solids*. John Wiley & Sons, March 2007.

[24] Ping He, Zeyi Tu, Guangyao Zhao, Yonggang Zhen, Hua Geng, Yuanping Yi, Zongrui Wang, Hantang Zhang, Chunhui Xu, Jie Liu, Xiuqiang Lu, Xiaolong Fu, Qiang Zhao, Xiaotao Zhang, Deyang Ji, Lang Jiang, Huanli Dong, and Wenping Hu. Tuning the Crystal Polymorphs of Alkyl Thienoacene via Solution Self-Assembly Toward Air-Stable and High-Performance Organic Field-Effect Transistors. *Advanced Materials*, 27(5):825–830, February 2015.

[25] Yu Yamashita, Felix Hinkel, Tomasz Marszalek, Wojciech Zajaczkowski, Wojciech Pisula, Martin Baumgarten, Hiroyuki Matsui, Klaus Müllen,

and Jun Takeya. Mobility Exceeding 10 cm2/(vs) in Donor-Acceptor Polymer Transistors with Band-like Charge Transport. *Chemistry of Materials*, 28(2):420–424, January 2016.

[26] J. Frenkel. On pre-breakdown phenomena in insulators and electronic semi-conductors. *Physical Review*, 54(8):647, 1938.

[27] C. D. Child. Discharge From Hot CaO. *Physical Review (Series I)*, 32(5):492–511, May 1911.

[28] H. Bässler. Charge Transport in Disordered Organic Photoconductors a Monte Carlo Simulation Study. *physica status solidi (b)*, 175(1):15–56, January 1993.

[29] Eric Mankel. *Elektronische Eigenschaften von Heterosystemen organischer und anorganischer Halbleiter: Präparation, Modifikation und Charakterisierung von Grenzflächen und Kompositen*. Dissertation, Technische Universität Darmstadt, 2011.

[30] Michael Kröger, Sami Hamwi, Jens Meyer, Thomas Riedl, Wolfgang Kowalsky, and Antoine Kahn. P-type doping of organic wide band gap materials by transition metal oxides: A case-study on Molybdenum trioxide. *Organic Electronics*, 10(5):932–938, August 2009.

[31] B. W. D'Andrade, R. J. Holmes, and S. R. Forrest. Efficient Organic Electrophosphorescent White-Light-Emitting Device with a Triple Doped Emissive Layer. *Advanced Materials*, 16(7):624–628, April 2004.

[32] T. Tsuzuki and S. Tokito. Highly Efficient and Low-Voltage Phosphorescent Organic Light-Emitting Diodes Using an Iridium Complex as the Host Material. *Advanced Materials*, 19(2):276–280, January 2007.

[33] Corinna Hein. *Anpassung der elektronischen Struktur an organischen Heterokontakten*. Dissertation, Technische Universität Darmstadt, 2012.

[34] Peter Würfel and Uli Würfel. *Physics of Solar Cells: From Basic Principles to Advanced Concepts*. John Wiley & Sons, March 2009.

[35] Koen Vandewal. Interfacial Charge Transfer States in Condensed Phase Systems. *Annual Review of Physical Chemistry*, 67(1):null, 2016.

[36] Bernard Kippelen and Jean-Luc Bredas. Organic photovoltaics. *Energy & Environmental Science*, 2(3):251–261, March 2009.

[37] Diana Nanova. *Intermolecular ordering in organic semiconductor layers and its correlation to electronic properties*. Dissertation, Universität Heidelberg, 2015.

[38] Andrius Devizis, Jelissa De Jonghe-Risse, Roland Hany, Frank Nüesch, Sandra Jenatsch, Vidmantas Gulbinas, and Jacques-E. Moser. Dissociation of Charge Transfer States and Carrier Separation in Bilayer Organic Solar Cells: A Time-Resolved Electroabsorption Spectroscopy Study. *Journal of the American Chemical Society*, 137(25):8192–8198, July 2015.

[39] Simon Gelinas, Akshay Rao, Abhishek Kumar, Samuel L. Smith, Alex W. Chin, Jenny Clark, Tom S. van der Poll, Guillermo C. Bazan, and Richard H. Friend. Ultrafast Long-Range Charge Separation in Organic Semiconductor Photovoltaic Diodes. *Science*, 343(6170):512–516, January 2014.

[40] Brett M. Savoie, Akshay Rao, Artem A. Bakulin, Simon Gelinas, Bijan Movaghar, Richard H. Friend, Tobin J. Marks, and Mark A. Ratner. Unequal Partnership: Asymmetric Roles of Polymeric Donor and Fullerene Acceptor in Generating Free Charge. *Journal of the American Chemical Society*, 136(7):2876–2884, February 2014.

[41] B. Bernardo, D. Cheyns, B. Verreet, R. D. Schaller, B. P. Rand, and N. C. Giebink. Delocalization and dielectric screening of charge transfer states in organic photovoltaic cells. *Nature Communications*, 5:3245, February 2014.

[42] Koen Vandewal, Steve Albrecht, Eric T. Hoke, Kenneth R. Graham, Johannes Widmer, Jessica D. Douglas, Marcel Schubert, William R. Mateker, Jason T. Bloking, George F. Burkhard, Alan Sellinger, Jean M. J. Frechet, Aram Amassian, Moritz K. Riede, Michael D. McGehee, Dieter Neher, and Alberto Salleo. Efficient charge generation by relaxed charge-transfer states at organic interfaces. *Nature Materials*, 13(1):63–68, January 2014.

[43] Johannes Widmer, Max Tietze, Karl Leo, and Moritz Riede. Open-Circuit Voltage and Effective Gap of Organic Solar Cells. *Advanced Functional Materials*, 23(46):5814–5821, December 2013.

[44] Thomas Kirchartz, Julian Mattheis, and Uwe Rau. Detailed balance theory of excitonic and bulk heterojunction solar cells. *Physical Review B*, 78(23):235320, December 2008.

[45] Andrew N. Bartynski, Mark Gruber, Saptaparna Das, Sylvie Rangan, Sonya Mollinger, Cong Trinh, Stephen E. Bradforth, Koen Vandewal, Alberto Salleo, Robert A. Bartynski, Wolfgang Bruetting, and Mark E. Thompson. Symmetry-Breaking Charge Transfer in a Zinc Chlorodipyrrin Acceptor for High Open Circuit Voltage Organic Photovoltaics. *Journal of the American Chemical Society*, 137(16):5397–5405, April 2015.

[46] Kjell Cnops, Barry P. Rand, David Cheyns, Bregt Verreet, Max A. Empl, and Paul Heremans. 8.4% efficient fullerene-free organic solar cells exploiting long-range exciton energy transfer. *Nature Communications*, 5:3406, March 2014.

[47] James Endres, Istvan Pelczer, Barry P. Rand, and Antoine Kahn. Determination of Energy Level Alignment within an Energy Cascade Organic Solar Cell. *Chemistry of Materials*, 28(3):794–801, February 2016.

[48] G. Yu, J. Gao, J. C. Hummelen, F. Wudl, and A. J. Heeger. Polymer Photovoltaic Cells: Enhanced Efficiencies via a Network of Internal Donor-Acceptor Heterojunctions. *Science*, 270(5243):1789–1791, December 1995.

[49] Harald Hoppe and Niyazi Serdar Sariciftci. Organic solar cells: An overview. *Journal of Materials Research*, 19(07):1924–1945, 2004.

[50] Boyuan Qi and Jizheng Wang. Open-circuit voltage in organic solar cells. *Journal of Materials Chemistry*, 22(46):24315, 2012.

[51] Thomas Kirchartz, Juan Bisquert, Ivan Mora-Sero, and Germa Garcia-Belmonte. Classification of solar cells according to mechanisms of charge

separation and charge collection. *Physical Chemistry Chemical Physics*, 17(6):4007–4014, January 2015.

[52] Brian A. Gregg and Mark C. Hanna. Comparing organic to inorganic photovoltaic cells: Theory, experiment, and simulation. *Journal of Applied Physics*, 93(6):3605–3614, March 2003.

[53] Gerold U. Bublitz and and Steven G. Boxer. STARK SPECTROSCOPY: Applications in Chemistry, Biology, and Materials Science. *Annual Review of Physical Chemistry*, 48(1):213–242, 1997.

[54] I. H. Campbell, T. W. Hagler, D. L. Smith, and J. P. Ferraris. Direct Measurement of Conjugated Polymer Electronic Excitation Energies Using Metal/Polymer/Metal Structures. *Physical Review Letters*, 76(11):1900–1903, March 1996.

[55] C. J. Brabec, A. Cravino, D. Meissner, N. S. Sariciftci, T. Fromherz, M. T. Rispens, L. Sanchez, and J. C. Hummelen. Origin of the Open Circuit Voltage of Plastic Solar Cells. *Advanced Functional Materials*, 11(5):374–380, October 2001.

[56] V. D. Mihailetchi, P. W. M. Blom, J. C. Hummelen, and M. T. Rispens. Cathode dependence of the open-circuit voltage of polymer:fullerene bulk heterojunction solar cells. *Journal of Applied Physics*, 94(10):6849–6854, November 2003.

[57] Ying-Xuan Wang, Shin-Rong Tseng, Hsin-Fei Meng, Kuan-Chen Lee, Chiou-Hua Liu, and Sheng-Fu Horng. Dark carrier recombination in organic solar cell. *Applied Physics Letters*, 93(13):133501, September 2008.

[58] Erin L. Ratcliff, Andres Garcia, Sergio A. Paniagua, Sarah R. Cowan, Anthony J. Giordano, David S. Ginley, Seth R. Marder, Joseph J. Berry, and Dana C. Olson. Investigating the Influence of Interfacial Contact Properties on Open Circuit Voltages in Organic Photovoltaic Performance: Work Function Versus Selectivity. *Advanced Energy Materials*, 3(5):647–656, May 2013.

[59] E. Siebert-Henze, V. G. Lyssenko, J. Fischer, M. Tietze, R. Brueckner, T. Menke, K. Leo, and M. Riede. Electroabsorption studies of organic p-i-n solar cells: Increase of the built-in voltage by higher doping concentration in the hole transport layer. *Organic Electronics*, 15(2):563–568, February 2014.

[60] Chetan R. Singh, Cheng Li, Christian J. Mueller, Sven Hüttner, and Mukundan Thelakkat. Influence of Electron Extracting Interface Layers in Organic Bulk-Heterojunction Solar Cells. *Advanced Materials Interfaces*, December 2015.

[61] Mi Zhou, Rui-Qi Png, Siong-Hee Khong, Sankaran Sivaramakrishnan, Li-Hong Zhao, Lay-Lay Chua, Richard H. Friend, and Peter K. H. Ho. Effective work functions for the evaporated metal/organic semiconductor contacts from in-situ diode flatband potential measurements. *Applied Physics Letters*, 101(1):013501, July 2012.

[62] G. Binnig, H. Rohrer, Ch. Gerber, and E. Weibel. Surface Studies by Scanning Tunneling Microscopy. *Physical Review Letters*, 49(1):57–61, July 1982.

[63] The Nobel prize in physics 1986. http://www.nobelprize.org/nobel_prizes/physics/laureates/1986/, April 2016.

[64] G. Binnig, C. F. Quate, and Ch. Gerber. Atomic Force Microscope. *Physical Review Letters*, 56(9):930–933, March 1986.

[65] Seizo Morita, Franz J. Giessibl, Ernst Meyer, and Roland Wiesendanger, editors. *Noncontact Atomic Force Microscopy*. NanoScience and Technology. Springer International Publishing, Cham, 2015.

[66] Hans-Jürgen Butt, Brunero Cappella, and Michael Kappl. Force measurements with the atomic force microscope: Technique, interpretation and applications. *Surface Science Reports*, 59(1-6):1–152, October 2005.

[67] Franz J. Giessibl. Advances in atomic force microscopy. *Reviews of Modern Physics*, 75(3):949–983, July 2003.

[68] Franz J. Giessibl. *Progress in atomic force microscopy*. Habilitation, Universität Augsburg, 2000.

[69] H. C. Hamaker. The London-van der Waals attraction between spherical particles. *Physica*, 4(10):1058–1072, October 1937.

[70] L. Olsson, N. Lin, V. Yakimov, and R. Erlandsson. A method for in situ characterization of tip shape in ac-mode atomic force microscopy using electrostatic interaction. *Journal of Applied Physics*, 84(8):4060–4064, October 1998.

[71] T. R. Albrecht, P. Grütter, D. Horne, and D. Rugar. Frequency modulation detection using high q factor cantilevers for enhanced force microscope sensitivity. *Journal of Applied Physics*, 69(2):668–673, January 1991.

[72] Q. Zhong, D. Inniss, K. Kjoller, and V. B. Elings. Fractured polymer/silica fiber surface studied by tapping mode atomic force microscopy. *Surface Science Letters*, 290(1):L688–L692, June 1993.

[73] Robert W. Carpick and Miquel Salmeron. Scratching the surface: fundamental investigations of tribology with atomic force microscopy. *Chemical Reviews*, 97(4):1163–1194, 1997.

[74] Dual Scope 95 from DME/Semilab. http://dme-spm.com/ds95.html, April 2016.

[75] Sebastian Hietzschold. *Vermessung der Zustandsdichte in organischen Feldeffekttransistoren durch Raster-Kelvin-Mikroskopie*. Master thesis, Universität Heidelberg, 2013.

[76] Data sheet Arrow NC from Nanoworld. http://www.nanoworld.com/tapping-mode-afm-tip-arrow-nc, April 2016.

[77] Janusz Schinke. *Processing and characterization of organic thin films: self-assembled monolayers for interface engineering and semiconductors with thermally activated solubility reduction*. Dissertation, Technische Universität Braunschweig, 2014.

[78] D. Cahen and A. Kahn. Electron Energetics at Surfaces and Interfaces: Concepts and Experiments. *Advanced Materials*, 15(4):271–277, February 2003.

[79] M. Nonnenmacher, M. P. O'Boyle, and H. K. Wickramasinghe. Kelvin probe force microscopy. *Applied Physics Letters*, 58(25):2921–2923, June 1991.

[80] Wilford N Hansen and Galen J Hansen. Standard reference surfaces for work function measurements in air. *Surface Science*, 481(1-3):172–184, June 2001.

[81] Peter Milde. *Visualization of local charge densities with Kelvin probe force microscopy*. Dissertation, Technische Universität Dresden, 2011.

[82] Ulrich Zerweck, Christian Loppacher, Tobias Otto, Stefan Grafström, and Lukas M. Eng. Accuracy and resolution limits of Kelvin probe force microscopy. *Physical Review B*, 71(12):125424, March 2005.

[83] Th Glatzel, S Sadewasser, and M. Ch Lux-Steiner. Amplitude or frequency modulation-detection in Kelvin probe force microscopy. *Applied Surface Science*, 210(1-2):84–89, March 2003.

[84] Prof. Angelika Kühnle. Personal communication, 2013.

[85] Shin'ichi Kitamura, Katsuyuki Suzuki, Masashi Iwatsuki, and C. B Mooney. Atomic-scale variations in contact potential difference on Au/Si(111) 7x7 surface in ultrahigh vacuum. *Applied Surface Science*, 157(4):222–227, April 2000.

[86] G. H. Enevoldsen, T. Glatzel, M. C. Christensen, J. V. Lauritsen, and F. Besenbacher. Atomic Scale Kelvin Probe Force Microscopy Studies of the Surface Potential Variations on the tio2 (110) Surface. *Physical Review Letters*, 100(23):236104, June 2008.

[87] Sascha Sadewasser, Pavel Jelinek, Chung-Kai Fang, Oscar Custance, Yusaku Yamada, Yoshiaki Sugimoto, Masayuki Abe, and Seizo Morita. New Insights on Atomic-Resolution Frequency-Modulation Kelvin-Probe Force-Microscopy Imaging of Semiconductors. *Physical Review Letters*, 103(26):266103, December 2009.

[88] Carmen Perez Leon, Holger Drees, Stefan Martin Wippermann, Michael Marz, and Regina Hoffmann-Vogel. Atomic-Scale Imaging of the Surface

Dipole Distribution of Stepped Surfaces. *The Journal of Physical Chemistry Letters*, pages 426–430, January 2016.

[89] Wilhelm Melitz, Jian Shen, Andrew C. Kummel, and Sangyeob Lee. Kelvin probe force microscopy and its application. *Surface Science Reports*, 66(1):1–27, January 2011.

[90] George Elias, Thilo Glatzel, Ernst Meyer, Alex Schwarzman, Amir Boag, and Yossi Rosenwaks. The role of the cantilever in Kelvin probe force microscopy measurements. *Beilstein Journal of Nanotechnology*, 2:252–260, May 2011.

[91] Dimitri S. H. Charrier, Martijn Kemerink, Barry E. Smalbrugge, Tjibbe de Vries, and Rene A. J. Janssen. Real versus Measured Surface Potentials in Scanning Kelvin Probe Microscopy. *ACS Nano*, 2(4):622–626, April 2008.

[92] E. Strassburg, A. Boag, and Y. Rosenwaks. Reconstruction of electrostatic force microscopy images. *Review of Scientific Instruments*, 76(8):083705, August 2005.

[93] T. Machleidt, E. Sparrer, D. Kapusi, and K.-H. Franke. Deconvolution of Kelvin probe force microscopy measurements- methodology and application. *Measurement Science and Technology*, 20(8):084017, 2009.

[94] Robert Baier. Toward quantitative Kelvin probe force microscopy of nanoscale potential distributions. *Physical Review B*, 85(16), 2012.

[95] Shengming Li, Yusheng Zhou, Yunlong Zi, Gong Zhang, and Zhong Lin Wang. Excluding Contact Electrification in Surface Potential Measurement Using Kelvin Probe Force Microscopy. *ACS Nano*, January 2016.

[96] Atsushi Kikukawa, Sumio Hosaka, and Ryo Imura. Silicon pn junction imaging and characterizations using sensitivity enhanced Kelvin probe force microscopy. *Applied Physics Letters*, 66(25):3510–3512, June 1995.

[97] Olivier Vatel and Masafumi Tanimoto. Kelvin probe force microscopy for potential distribution measurement of semiconductor devices. *Journal of Applied Physics*, 77(6):2358–2362, March 1995.

Bibliography

[98] A. Chavez-Pirson, O. Vatel, M. Tanimoto, H. Ando, H. Iwamura, and H. Kanbe. Nanometer scale imaging of potential profiles in optically excited nipi heterostructure using kelvin probe force microscopy. *Applied Physics Letters*, 67(21):3069–3071, November 1995.

[99] Data sheet ATEC from Nanosensors. http://goo.gl/zeVUGQ, April 2016.

[100] D. McMullan. Scanning electron microscopy 1928-1965. *Scanning*, 17(3):175–185, May 1995.

[101] T. E. Everhart and R. F. M. Thornley. Wide-band detector for micro-microampere low-energy electron currents. *Journal of Scientific Instruments*, 37(7):246, 1960.

[102] Goerg H. Michler. *Electron Microscopy of Polymers*. Springer Science & Business Media, July 2008.

[103] W. H. Escovitz, T. R. Fox, and R. Levi-Setti. Scanning transmission ion microscope with a field ion source. *Proceedings of the National Academy of Sciences*, 72(5):1826–1828, January 1975.

[104] J. H. Orloff and L. W. Swanson. Study of a field ionization source for microprobe applications. *Journal of Vacuum Science & Technology*, 12(6):1209–1213, November 1975.

[105] C. A. Volkert and A. M. Minor. Focused Ion Beam Microscopy and Micromachining. *MRS Bulletin*, 32(05):389–399, May 2007.

[106] John A. Notte. Charged Particle Microscopy: Why Mass Matters. *Microscopy Today*, 20(05):16–22, 2012.

[107] Vadim Sidorkin, Emile van Veldhoven, Emile van der Drift, Paul Alkemade, Huub Salemink, and Diederik Maas. Sub-10-nm nanolithography with a scanning helium beam. *Journal of Vacuum Science & Technology B*, 27(4):L18–L20, July 2009.

[108] Donald Winston, Vitor R. Manfrinato, Samuel M. Nicaise, Lin Lee Cheong, Huigao Duan, David Ferranti, Jeff Marshman, Shawn McVey, Lewis Stern, John Notte, and Karl K. Berggren. Neon Ion Beam Lithography (NIBL). *Nano Letters*, 11(10):4343–4347, October 2011.

[109] Ludovit P. Zweifel, Ivan Shorubalko, and Roderick Y.H. Lim. Helium Scanning Transmission Ion Microscopy and Electrical Characterization of Glass Nanocapillaries with Reproducible Tip Geometries. *ACS Nano*, January 2016.

[110] Gregor Hlawacek, Vasilisa Veligura, Raoul van Gastel, and Bene Poelsema. Helium ion microscopy. *Journal of Vacuum Science & Technology B*, 32(2):020801, March 2014.

[111] D. Elswick, M. Ananth, L. Stern, J. Marshman, D. Ferranti, and C. Huynh. Advanced Nanofabrication using Helium, Neon and Gallium Ion Beams in the Carl Zeiss Orion NanoFab Microscope. *Microscopy and Microanalysis*, 19(Supplement S2):1304–1305, August 2013.

[112] Zeiss Orion NanoFab. http://www.zeiss.com/microscopy/en_de/products/multiple-ion-beam/orion-nanofab-for-materials.html, April 2016.

[113] Noriaki Matsunami, Yasunori Yamamura, Yukikazu Itikawa, Noriaki Itoh, Yukio Kazumata, Soji Miyagawa, Kenji Morita, Ryuichi Shimizu, and Hiroyuki Tawara. Energy dependence of the ion-induced sputtering yields of monatomic solids. *Atomic Data and Nuclear Data Tables*, 31(1):1–80, July 1984.

[114] Yaunori Yamamura and Hiro Tawara. Energy dependence og ion-induced sputtering yields from monatomic solids at normal incidence. *Atomic Data and Nuclear Data Tables*, 62(2):149–253, March 1996.

[115] Homepage of the National Physical Laboratory London. http://www.npl.co.uk/science-technology/surface-and-nanoanalysis/services/sputter-yield-values, April 2016.

[116] M. P. Seah. An accurate semi-empirical equation for sputtering yields, II: for neon, argon and xenon ions. *Nuclear Instruments and Methods in Physics Research Section B: Beam Interactions with Materials and Atoms*, 229(3-4):348–358, April 2005.

[117] Hu Li, Kazuhiro Karahashi, Masanaga Fukasawa, Kazunori Nagahata, Tetsuya Tatsumi, and Satoshi Hamaguchi. Sputtering yields and sur-

face chemical modification of tin-doped indium oxide in hydrocarbon-based plasma etching. *Journal of Vacuum Science & Technology A*, 33(6):060606, November 2015.

[118] BRR AFM-SEM integration from DME/Semilab. http://www.dme-spm.com/remafm.html, April 2016.

[119] Rebecca Saive. *Investigation of the potential distribution within organic solar cells by scanning Kelvin probe microscopy*. Dissertation, Universität Heidelberg, 2014.

[120] David B. Williams and C. Barry Carter. High-Resolution TEM. In *Transmission Electron Microscopy*, pages 483–509. Springer US, 2009. DOI: 10.1007/978-0-387-76501-3_28.

[121] Martin Pfannmöller. *Visualisierung von Nanostrukturen in organischen, photoaktiven Mischsystemen durch dreidimensionale analytische Elektronenmikroskopie*. Dissertation, Universität Heidelberg, 2013.

[122] Martin Pfannmöller, Harald Flügge, Gerd Benner, Irene Wacker, Christoph Sommer, Michael Hanselmann, Stephan Schmale, Hans Schmidt, Fred A. Hamprecht, Torsten Rabe, Wolfgang Kowalsky, and Rasmus R. Schröder. Visualizing a Homogeneous Blend in Bulk Heterojunction Polymer Solar Cells by Analytical Electron Microscopy. *Nano Letters*, 11(8):3099–3107, August 2011.

[123] Diana Nanova, Anne Katrin Kast, Martin Pfannmöller, Christian Müller, Lisa Veith, Irene Wacker, Michaela Agari, Wilfried Hermes, Peter Erk, Wolfgang Kowalsky, Rasmus R. Schröder, and Robert Lovrincic. Unraveling the Nanoscale Morphologies of Mesoporous Perovskite Solar Cells and Their Correlation to Device Performance. *Nano Letters*, 14(5):2735–2740, May 2014.

[124] Julia Maibach. *Preparation and characterization of solution-processed organic semiconductor interfaces: electronic properties of thiophene-fullerene based donor-acceptor systems*. Dissertation, Technische Universität Darmstadt, 2014.

[125] Eric Mankel. *Grundlagen der Röntgenphotoelektronenspektroskopie, Praktikumsversuch Bachelor 5. Semester, WS 2011/2012.* FG Oberflächenforschung, Technische Universität Darmstadt, 2011.

[126] Andreas Klein, Thomas Mayer, Andreas Thissen, and Wolfram Jaegermann. Photoelectron Spectroscopy in Materials Science and Physical Chemistry: Analysis of Composition, Chemical Bonding, and Electronic Structure of Surfaces and Interfaces. In Rolf Schäfer and Peter C. Schmidt, editors, *Methods in Physical Chemistry*, pages 477–512. Wiley-VCH Verlag GmbH & Co. KGaA, 2012.

[127] A. Einstein. Über einen die Erzeugung und Verwandlung des Lichtes betreffenden heuristischen Gesichtspunkt [AdP 17, 132 (1905)]. *Annalen der Physik*, 14(S1):164–181, February 2005.

[128] Ilja Vladimirov. *Entwicklung eines Nahfeld-Rastersondenmikroskops auf Grundlage organischer Leuchtdioden.* Diploma thesis, Universität Heidelberg, 2011.

[129] Benjamin Martini. *Studien zur Realisierung einer nanoskopischen organischen Leuchtdiode auf der Spitze einer Rasterkraftsonde.* Diploma thesis, Technische Universität München, 2012.

[130] Paul Heimel. *Einfluss einer MoO3-Dotierung in CBP auf die Bauteileigenschaften von OLEDs.* Master thesis, Universität Heidelberg, 2013.

[131] E. Putseiko. Primenenie metoda kondensatora k kizucheniyu vnutrennego fotoeffekta sensibilzatorov. *Doklady Akademii Nauk SSSR*, 59(3):471–474, 1948.

[132] Andreas Opitz, Michael Kraus, Markus Bronner, Julia Wagner, and Wolfgang Brütting. Bipolar transport in organic field-effect transistors: organic semiconductor blends versus contact modification. *New Journal of Physics*, 10(6):065006, 2008.

[133] Andreas Opitz, Julia Wagner, Wolfgang Brütting, Alexander Hinderhofer, and Frank Schreiber. Molecular semiconductor blends: Microstructure, charge carrier transport, and application in photovoltaic cells. *physica status solidi (a)*, 206(12):2683–2694, December 2009.

[134] Steffen Pfuetzner, Christine Mickel, Jens Jankowski, Moritz Hein, Jan Meiss, Christoph Schuenemann, Chris Elschner, Alexandr A. Levin, Bernd Rellinghaus, Karl Leo, and Moritz Riede. The influence of substrate heating on morphology and layer growth in C60:ZnPc bulk heterojunction solar cells. *Organic Electronics*, 12(3):435–441, March 2011.

[135] N. El Khatib, B. Boudjema, M. Maitrot, H. Chermette, and L. Porte. Electronic structure of zinc phthalocyanine. *Canadian Journal of Chemistry*, 66(9):2313–2324, September 1988.

[136] C. Schünemann, C. Elschner, A. A. Levin, M. Levichkova, K. Leo, and M. Riede. Zinc phthalocyanine - Influence of substrate temperature, film thickness, and kind of substrate on the morphology. *Thin Solid Films*, 519(11):3939–3945, March 2011.

[137] K. Itaka, M. Yamashiro, J. Yamaguchi, M. Haemori, S. Yaginuma, Y. Matsumoto, M. Kondo, and H. Koinuma. High-Mobility C60 Field-Effect Transistors Fabricated on Molecular- Wetting Controlled Substrates. *Advanced Materials*, 18(13):1713–1716, July 2006.

[138] D. Faiman, S. Goren, E. A. Katz, M. Koltun, N. Melnik, A. Shames, and S. Shtutina. Structure and optical properties of C60 thin films. *Thin Solid Films*, 295(1-2):283–286, February 1997.

[139] Fu-Zhou Sun, Ai-Li Shi, Zai-Quan Xu, Huai-Xin Wei, Yan-Qing Li, Shuit-Tong Lee, and Jian-Xin Tang. Efficient inverted polymer solar cells with thermal-evaporated and solution-processed small molecular electron extraction layer. *Applied Physics Letters*, 102(13):133303, April 2013.

[140] Vibha Tripathi, Debjit Datta, G. S. Samal, Asha Awasthi, and Satyendra Kumar. Role of exciton blocking layers in improving efficiency of copper phthalocyanine based organic solar cells. *Journal of Non-Crystalline Solids*, 354(19-25):2901–2904, May 2008.

[141] Junsheng Yu, Nana Wang, Yue Zang, and Yadong Jiang. Organic photovoltaic cells based on TPBi as a cathode buffer layer. *Solar Energy Materials and Solar Cells*, 95(2):664–668, February 2011.

[142] Jens Meyer, Sami Hamwi, Michael Kröger, Wolfgang Kowalsky, Thomas Riedl, and Antoine Kahn. Transition Metal Oxides for Organic Electronics: Energetics, Device Physics and Applications. *Advanced Materials*, 24(40):5408–5427, October 2012.

[143] Yanming Sun, Christopher J. Takacs, Sarah R. Cowan, Jung Hwa Seo, Xiong Gong, Anshuman Roy, and Alan J. Heeger. Efficient, Air-Stable Bulk Heterojunction Polymer Solar Cells Using MoOx as the Anode Interfacial Layer. *Advanced Materials*, 23(19):2226–2230, May 2011.

[144] Philip Schulz, Sarah R. Cowan, Ze-Lei Guan, Andres Garcia, Dana C. Olson, and Antoine Kahn. NiOX/MoO3 Bi-Layers as Efficient Hole Extraction Contacts in Organic Solar Cells. *Advanced Functional Materials*, 24(5):701–706, February 2014.

[145] Daniela Donhauser. *Correlation between structural and electronic properties of co-evaporated doped organic thin films*. Dissertation, Technische Universität Braunschweig, 2013.

[146] Tobias Glaser. *Infrarotspektroskopische Untersuchung der p-Dotierung organischer Halbleiter mit Übergangsmetalloxiden*. Dissertation, Universität Heidelberg, 2013.

[147] Sebastian Beck. *Untersuchung des Ladungstransfers in organischen Halbleitern mit in-situ Infrarotspektroskopie*. Dissertation, Universität Heidelberg, 2014.

[148] Maybritt Kühn. *Korrelation morphologischer und elektronischer Eigenschaften von dotierten organischen Halbleitersystemen*. Diploma thesis, Technische Universität Darmstadt, 2012.

[149] Jesse R. Manders, Sai-Wing Tsang, Michael J. Hartel, Tzung-Han Lai, Song Chen, Chad M. Amb, John R. Reynolds, and Franky So. Solution-Processed Nickel Oxide Hole Transport Layers in High Efficiency Polymer Photovoltaic Cells. *Advanced Functional Materials*, 23(23):2993–3001, June 2013.

[150] Kuo-Chin Wang, Jun-Yuan Jeng, Po-Shen Shen, Yu-Cheng Chang, Eric Wei-Guang Diau, Cheng-Hung Tsai, Tzu-Yang Chao, Hsu-Cheng Hsu,

Pei-Ying Lin, Peter Chen, Tzung-Fang Guo, and Ten-Chin Wen. p-type Mesoscopic Nickel Oxide/Organometallic Perovskite Heterojunction Solar Cells. *Scientific Reports*, 4, April 2014.

[151] Yinhua Zhou, Canek Fuentes-Hernandez, Jaewon Shim, Jens Meyer, Anthony J. Giordano, Hong Li, Paul Winget, Theodoros Papadopoulos, Hyeunseok Cheun, Jungbae Kim, Mathieu Fenoll, Amir Dindar, Wojciech Haske, Ehsan Najafabadi, Talha M. Khan, Hossein Sojoudi, Stephen Barlow, Samuel Graham, Jean-Luc Bredas, Seth R. Marder, Antoine Kahn, and Bernard Kippelen. A Universal Method to Produce Low-Work Function Electrodes for Organic Electronics. *Science*, 336(6079):327–332, April 2012.

[152] Felix Schell. *Morphology of small molecule organic solar cells from transmission electron microscopy*. Master thesis, Universität Heidelberg, 2014.

[153] Alexander Müller-Brand. *Development of a cryostat system and its subsequent operation in temperature-dependent conductivity measurements on MoO3-doped CBP*. Master thesis, Universität Heidelberg, 2013.

[154] Andreas Opitz, Andreas Wilke, Patrick Amsalem, Martin Oehzelt, Ralf-Peter Blum, Jürgen P. Rabe, Toshiko Mizokuro, Ulrich Hörmann, Rickard Hansson, Ellen Moons, and Norbert Koch. Organic heterojunctions: Contact-induced molecular reorientation, interface states, and charge re-distribution. *Scientific Reports*, 6, February 2016.

[155] Zhanhao Hu, Zhiming Zhong, Yawen Chen, Chen Sun, Fei Huang, Junbiao Peng, Jian Wang, and Yong Cao. Energy-Level Alignment at the Organic/Electrode Interface in Organic Optoelectronic Devices. *Advanced Functional Materials*, 26(1):129–136, January 2016.

[156] Masahiro Hiramoto, Keiji Koyama, Ken-ichi Nakayama, and Masaaki Yokoyama. Direct measurement of internal potential distribution in organic electroluminescent diodes during operation. *Applied Physics Letters*, 76(10):1336–1338, March 2000.

[157] X. R. Yin, Y. K. Le, X. D. Gao, Z. Y. Sun, and X. Y. Hou. Internal

potential distribution in organic light emitting diodes measured by dc bridge. *Applied Physics Letters*, 97(15):153305, October 2010.

[158] J. Meyer, A. Shu, M. Kröger, and A. Kahn. Effect of contamination on the electronic structure and hole-injection properties of MoO3/organic semiconductor interfaces. *Applied Physics Letters*, 96(13):133308, March 2010.

[159] Sami Hamwi. *Transition metal oxides in organic light emitting diodes*. Dissertation, Technische Universität Braunschweig, 2010.

[160] Michael Brendel, Stefan Krause, Andreas Steindamm, Anna Katharina Topczak, Sudhakar Sundarraj, Peter Erk, Steffen Höhla, Norbert Fruehauf, Norbert Koch, and Jens Pflaum. The Effect of Gradual Fluorination on the Properties of FnZnPc Thin Films and FnZnPc/C60 Bilayer Photovoltaic Cells. *Advanced Functional Materials*, 25(10):1565–1573, March 2015.

[161] M. Chelvayohan and C. H. B. Mee. Work function measurements on (110), (100) and (111) surfaces of silver. *Journal of Physics C: Solid State Physics*, 15(10):2305, 1982.

[162] Corinna Hein, Eric Mankel, Thomas Mayer, and Wolfram Jaegermann. Engineering the electronic structure of the CuPc/BPE-PTCDI interface by WO3 doping of CuPc. *physica status solidi (a)*, 206(12):2757–2762, December 2009.

[163] Corinna Hein, Eric Mankel, Thomas Mayer, and Wolfram Jaegermann. Engineering the electronic structure of the ZnPc/C60 heterojunction by temperature treatment. *Solar Energy Materials and Solar Cells*, 94(4):662–667, April 2010.

[164] Hiroyuki Yoshida. Electron Transport in Bathocuproine Interlayer in Organic Semiconductor Devices. *The Journal of Physical Chemistry C*, 119(43):24459–24464, October 2015.

[165] Y. Park, V. Choong, Y. Gao, B. R. Hsieh, and C. W. Tang. Work function of indium tin oxide transparent conductor measured by photoelectron spectroscopy. *Applied Physics Letters*, 68(19):2699–2701, May 1996.

[166] A. Wagenpfahl. S-shaped current-voltage characteristics of organic solar devices. *Physical Review B*, 82(11), 2010.

[167] Yvonne Gassenbauer and Andreas Klein. Electronic and Chemical Properties of Tin-Doped Indium Oxide (ITO) Surfaces and ITO/ZnPc Interfaces Studied In-situ by Photoelectron Spectroscopy. *The Journal of Physical Chemistry B*, 110(10):4793–4801, March 2006.

[168] J. Piekoszewski, L. Castaner, J. J. Loferski, J. Beall, and W. Giriat. Measurement of diffusion length in CuInSe2 and CdS by the electron beam induced current method. *Journal of Applied Physics*, 51(10):5375–5379, October 1980.

[169] H. J. Leamy. Charge collection scanning electron microscopy. *Journal of Applied Physics*, 53(6):R51–R80, June 1982.

[170] C. Ballif, H. R. Moutinho, and M. M. Al-Jassim. Cross-sectional electrostatic force microscopy of thin-film solar cells. *Journal of Applied Physics*, 89(2):1418–1424, January 2001.

[171] C. S. Jiang, H. R. Moutinho, M. J. Romero, M. M. Al-Jassim, Y. Q. Xu, and Q. Wang. Distribution of the electrical potential in hydrogenated amorphous silicon solar cells. *Thin Solid Films*, 472(1-2):203–207, January 2005.

[172] Rebecca Saive, Lars Mueller, Eric Mankel, Wolfgang Kowalsky, and Michael Kroeger. Doping of TIPS-pentacene via Focused Ion Beam (FIB) exposure. *Organic Electronics*, 14(6):1570–1576, June 2013.

[173] Michael Scherer, Rebecca Saive, Dominik Daume, Michael Kröger, and Wolfgang Kowalsky. Sample preparation for scanning Kelvin probe microscopy studies on cross sections of organic solar cells. *AIP Advances*, 3(9):092134, September 2013.

[174] Christian S. Weigel, Wolfgang Kowalsky, and Rebecca Saive. Direct observation of the potential distribution within organic light emitting diodes under operation. *physica status solidi (RRL) - Rapid Research Letters*, 9(8):475–479, August 2015.

[175] Tobias Jenne. *Raster-Kelvin-Mikroskopie an Querschnitten organischer F4ZnPc/C60-Solarzellen*. Master thesis, Universität Heidelberg, 2014.

[176] J. F. Gibbons. Ion implantation in semiconductors.Part I: Range distribution theory and experiments. *Proceedings of the IEEE*, 56(3):295–319, March 1968.

[177] W. D. Cussins. Effects Produced by the Ionic Bombardment of Germanium. *Proceedings of the Physical Society. Section B*, 68(4):213, 1955.

[178] F. M. Rourke, J. C. Sheffield, and F. A. White. Crystal "Doping" by Ion Bombardment. *Rev. Sci. Instr.*, Vol: 32, April 1961.

[179] J. O. McCaldin and A. E. Widmer. Silicon heavily doped by energetic cesium ions. *Journal of Physics and Chemistry of Solids*, 24(9):1073–1080, September 1963.

[180] Yoshiro Hirayama and Hiroshi Okamoto. Electrical Properties of Ga Ion Beam Implanted GaAs Epilayer. *Japanese Journal of Applied Physics*, 24(Part 2, No. 12):L965–L967, December 1985.

[181] T. Burchhart, C. Zeiner, A. Lugstein, C. Henkel, and E. Bertagnolli. Tuning the electrical performance of Ge nanowire MOSFETs by focused ion beam implantation. *Nanotechnology*, 22(3):035201, 2011.

[182] Gaurav Nanda, Srijit Goswami, Kenji Watanabe, Takashi Taniguchi, and Paul F. A. Alkemade. Defect Control and n-Doping of Encapsulated Graphene by Helium-Ion-Beam Irradiation. *Nano Letters*, 15(6):4006–4012, June 2015.

[183] Daniel S. Fox, Yangbo Zhou, Pierce Maguire, Arlene O'Neill, Cormac O'Coileain, Riley Gatensby, Alexey M. Glushenkov, Tao Tao, Georg S. Duesberg, Igor V. Shvets, Mohamed Abid, Mourad Abid, Han-Chun Wu, Ying Chen, Jonathan N. Coleman, John F. Donegan, and Hongzhou Zhang. Nanopatterning and Electrical Tuning of MoS2 Layers with a Subnanometer Helium Ion Beam. *Nano Letters*, 15(8):5307–5313, August 2015.

[184] Vighter Iberi, Anton V. Ievlev, Ivan Vlassiouk, Stephen Jesse, Sergei V. Kalinin, David C. Joy, Adam J. Rondinone, Alex Belianinov, and Olga S. Ovchinnikova. Graphene engineering by neon ion beams. *Nanotechnology*, 27(12):125302, 2016.

[185] James F. Ziegler, M. D. Ziegler, and J. P. Biersack. SRIM - The stopping and range of ions in matter (2010). *Nuclear Instruments and Methods in Physics Research Section B: Beam Interactions with Materials and Atoms*, 268(11-12):1818–1823, June 2010.

[186] R. Shikler, T. Meoded, N. Fried, and Y. Rosenwaks. Potential imaging of operating light-emitting devices using Kelvin force microscopy. *Applied Physics Letters*, 74(20):2972–2974, May 1999.

[187] C.-S. Jiang, H. R. Moutinho, R. Reedy, M. M. Al-Jassim, and A. Blosse. Two-dimensional junction identification in multicrystalline silicon solar cells by scanning Kelvin probe force microscopy. *Journal of Applied Physics*, 104(10):104501, November 2008.

[188] A Breymesser, V Schlosser, D Peiro, C Voz, J Bertomeu, J Andreu, and J Summhammer. Kelvin probe measurements of microcrystalline silicon on a nanometer scale using SFM. *Solar Energy Materials and Solar Cells*, 66(1-4):171–177, February 2001.

[189] H. R. Moutinho, R. G. Dhere, C.-S. Jiang, Yanfa Yan, D. S. Albin, and M. M. Al-Jassim. Investigation of potential and electric field profiles in cross sections of CdTe/CdS solar cells using scanning Kelvin probe microscopy. *Journal of Applied Physics*, 108(7):074503, October 2010.

[190] Zhenhao Zhang, Michael Hetterich, Uli Lemmer, Michael Powalla, and Hendrik Hölscher. Cross sections of operating Cu(In,Ga)Se2 thin-film solar cells under defined white light illumination analyzed by Kelvin probe force microscopy. *Applied Physics Letters*, 102(2):023903, January 2013.

[191] Victor W. Bergmann, Stefan A. L. Weber, F. Javier Ramos, Mohammad Khaja Nazeeruddin, Michael Grätzel, Dan Li, Anna L. Domanski,

Ingo Lieberwirth, Shahzada Ahmad, and Rüdiger Berger. Real-space observation of unbalanced charge distribution inside a perovskite-sensitized solar cell. *Nature Communications*, 5:5001, September 2014.

[192] Chun-Sheng Jiang, H. R. Moutinho, D. J. Friedman, J. F. Geisz, and M. M. Al-Jassim. Measurement of built-in electrical potential in III-V solar cells by scanning Kelvin probe microscopy. *Journal of Applied Physics*, 93(12):10035–10040, June 2003.

[193] Jongjin Lee, Jaemin Kong, Heejoo Kim, Sung-Oong Kang, and Kwanghee Lee. Direct observation of internal potential distributions in a bulk heterojunction solar cell. *Applied Physics Letters*, 99(24):243301, December 2011.

[194] Jaemin Kong, Jongjin Lee, Yonkil Jeong, Maengjun Kim, Sung-Oong Kang, and Kwanghee Lee. Biased internal potential distributions in a bulk-heterojunction organic solar cell incorporated with a TiOx interlayer. *Applied Physics Letters*, 100(21):213305, May 2012.

[195] L. Bürgi, H. Sirringhaus, and R. H. Friend. Noncontact potentiometry of polymer field-effect transistors. *Applied Physics Letters*, 80(16):2913–2915, April 2002.

[196] Qi Chen, Lin Mao, Yaowen Li, Tao Kong, Na Wu, Changqi Ma, Sai Bai, Yizheng Jin, Dan Wu, Wei Lu, Bing Wang, and Liwei Chen. Quantitative operando visualization of the energy band depth profile in solar cells. *Nature Communications*, 6:7745, July 2015.

[197] Edsger C. P. Smits, Simon G. J. Mathijssen, Michael Colle, Arjan J. G. Mank, Peter A. Bobbert, Paul W. M. Blom, Bert de Boer, and Dago M. de Leeuw. Unified description of potential profiles and electrical transport in unipolar and ambipolar organic field-effect transistors. *Physical Review B*, 76(12):125202, September 2007.

[198] Kanan P. Puntambekar, Paul V. Pesavento, and C. Daniel Frisbie. Surface potential profiling and contact resistance measurements on operating pentacene thin-film transistors by Kelvin probe force microscopy. *Applied Physics Letters*, 83(26):5539–5541, December 2003.

Bibliography

[199] Hagen Klauk, Günter Schmid, Wolfgang Radlik, Werner Weber, Lisong Zhou, Chris D Sheraw, Jonathan A Nichols, and Thomas N Jackson. Contact resistance in organic thin film transistors. *Solid-State Electronics*, 47(2):297–301, February 2003.

[200] Milan Alt, Malte Jesper, Janusz Schinke, Sabina Hillebrandt, Patrick Reiser, Tobias Rödlmeier, Iva Angelova, Kaja Deing, Tobias Glaser, Eric Mankel, Wolfram Jaegermann, Annemarie Pucci, Uli Lemmer, U. H. F. Bunz, W. Kowalsky, G. Hernandez-Sosa, R. Lovrincic, and M. Hamburger. The Swiss-Army-Knife Self-Assembled Monolayer: Improving Electron Injection, Stability, and Wettability of Metal Electrodes with a One-Minute Proces. *Advanced Functional Materials*, pages n/a–n/a, February 2016.

[201] Johannes Zimmermann. *Eine infrarotspektroskopische Studie an organischen Materialien bezüglich molekularer Orientierung und dem Nachweis chemischer Reaktionen*. Master thesis, Universität Heidelberg, 2014.

[202] Eric Mankel, Corinna Hein, Maybritt Kühn, and Thomas Mayer. Electric potential distributions in space charge regions of molecular organic adsorbates using a simplified distributed states model. *physica status solidi (a)*, 211(9):2040–2048, September 2014.

[203] Johannes Widmer, Janine Fischer, Wolfgang Tress, Karl Leo, and Moritz Riede. Electric potential mapping by thickness variation: A new method for model-free mobility determination in organic semiconductor thin films. *Organic Electronics*, 14(12):3460–3471, December 2013.

[204] J. J. M. Halls, K. Pichler, R. H. Friend, S. C. Moratti, and A. B. Holmes. Exciton diffusion and dissociation in a poly(p-phenylenevinylene)/C60 heterojunction photovoltaic cell. *Applied Physics Letters*, 68(22):3120–3122, May 1996.

[205] Takaaki Manaka and Mitsumasa Iwamoto. Electrical properties of unsubstituted/fluorine-substituted phthalocyanine interface investigated by Kelvin probe method. *Thin Solid Films*, 438-439:157–161, August 2003.

[206] Maojie Zhang, Xia Guo, Shaoqing Zhang, and Jianhui Hou. Synergistic Effect of Fluorination on Molecular Energy Level Modulation in Highly Efficient Photovoltaic Polymers. *Advanced Materials*, 26(7):1118–1123, February 2014.

[207] Claudio Girotto, Eszter Voroshazi, David Cheyns, Paul Heremans, and Barry P. Rand. Solution-Processed MoO3 Thin Films As a Hole-Injection Layer for Organic Solar Cells. *ACS Applied Materials & Interfaces*, 3(9):3244–3247, September 2011.

[208] V. Kazukauskas, A. Arlauskas, M. Pranaitis, R. Lessmann, M. Riede, and K. Leo. Conductivity, charge carrier mobility and ageing of ZnPc/C60 solar cells. *Optical Materials*, 32(12):1676–1680, October 2010.

[209] Brian E. Lassiter, Guodan Wei, Siyi Wang, Jeramy D. Zimmerman, Viacheslav V. Diev, Mark E. Thompson, and Stephen R. Forrest. Organic photovoltaics incorporating electron conducting exciton blocking layers. *Applied Physics Letters*, 98(24):243307, June 2011.

[210] Andrew N. Bartynski, Cong Trinh, Anurag Panda, Kevin Bergemann, Brian E. Lassiter, Jeramy D. Zimmerman, Stephen R. Forrest, and Mark E. Thompson. A Fullerene-Based Organic Exciton Blocking Layer with High Electron Conductivity. *Nano Letters*, 13(7):3315–3320, July 2013.

[211] Wolfgang Tress, Karl Leo, and Moritz Riede. Influence of Hole-Transport Layers and Donor Materials on Open-Circuit Voltage and Shape of I-V Curves of Organic Solar Cells. *Advanced Functional Materials*, 21(11):2140–2149, June 2011.

[212] Hong Li, Paul Winget, and Jean-Luc Bredas. Transparent Conducting Oxides of Relevance to Organic Electronics: Electronic Structures of Their Interfaces with Organic Layers. *Chemistry of Materials*, 26(1):631–646, January 2014.

[213] K. Xerxes Steirer, Jordan P. Chesin, N. Edwin Widjonarko, Joseph J. Berry, Alex Miedaner, David S. Ginley, and Dana C. Olson. Solution

deposited NiO thin-films as hole transport layers in organic photovoltaics. *Organic Electronics*, 11(8):1414–1418, August 2010.

[214] Erin L. Ratcliff, Jens Meyer, K. Xerxes Steirer, Neal R. Armstrong, Dana Olson, and Antoine Kahn. Energy level alignment in PCDTBT:PC70bm solar cells: Solution processed NiOx for improved hole collection and efficiency. *Organic Electronics*, 13(5):744–749, May 2012.

[215] H. H. P. Gommans, D. Cheyns, T. Aernouts, C. Girotto, J. Poortmans, and P. Heremans. Electro-Optical Study of Subphthalocyanine in a Bilayer Organic Solar Cell. *Advanced Functional Materials*, 17(15):2653–2658, October 2007.

[216] Barry P. Rand, David Cheyns, Karolien Vasseur, Noel C. Giebink, Sebastien Mothy, Yuanping Yi, Veaceslav Coropceanu, David Beljonne, Jerome Cornil, Jean-Luc Bredas, and Jan Genoe. The Impact of Molecular Orientation on the Photovoltaic Properties of a Phthalocyanine/Fullerene Heterojunction. *Advanced Functional Materials*, 22(14):2987–2995, July 2012.

[217] Bregt Verreet, Robert Müller, Barry P. Rand, Karolien Vasseur, and Paul Heremans. Structural templating of chloro-aluminum phthalocyanine layers for planar and bulk heterojunction organic solar cells. *Organic Electronics*, 12(12):2131–2139, December 2011.

[218] Ulrich Hörmann, Christopher Lorch, Alexander Hinderhofer, Alexander Gerlach, Mark Gruber, Julia Kraus, Benedikt Sykora, Stefan Grob, Theresa Linderl, Andreas Wilke, Andreas Opitz, Rickard Hansson, Ana Sofia Anselmo, Yusuke Ozawa, Yasuo Nakayama, Hisao Ishii, Norbert Koch, Ellen Moons, Frank Schreiber, and Wolfgang Brütting. Voc from a Morphology Point of View: the Influence of Molecular Orientation on the Open Circuit Voltage of Organic Planar Heterojunction Solar Cells. *The Journal of Physical Chemistry C*, 118(46):26462–26470, November 2014.

[219] Julia L. Neff, Peter Milde, Carmen Perez Leon, Matthew D. Kundrat, Lukas M. Eng, Christoph R. Jacob, and Regina Hoffmann-Vogel. Epitaxial Growth of Pentacene on Alkali Halide Surfaces Studied by Kelvin Probe Force Microscopy. *ACS Nano*, 8(4):3294–3301, April 2014.

[220] Franz Fuchs, Florent Caffy, Renaud Demadrille, Thierry Melin, and Benjamin Grevin. High-Resolution Kelvin Probe Force Microscopy Imaging of Interface Dipoles and Photogenerated Charges in Organic Donor-Acceptor Photovoltaic Blends. *ACS Nano*, 10(1):739–746, January 2016.

[221] Berkem Ozkaya, Simon Grosse-Kreul, Carles Corbella, Achim von Keudell, and Guido Grundmeier. Combined In Situ XPS and UHV-Chemical Force Microscopy (CFM) Studies of the Plasma Induced Surface Oxidation of Polypropylene. *Plasma Processes and Polymers*, 11(3):256–262, March 2014.

[222] Yoshihiro Kanai, Toshinori Matsushima, and Hideyuki Murata. Improvement of stability for organic solar cells by using molybdenum trioxide buffer layer. *Thin Solid Films*, 518(2):537–540, November 2009.

[223] Sung Heum Park, Anshuman Roy, Serge Beaupre, Shinuk Cho, Nelson Coates, Ji Sun Moon, Daniel Moses, Mario Leclerc, Kwanghee Lee, and Alan J. Heeger. Bulk heterojunction solar cells with internal quantum efficiency approaching 100%. *Nature Photonics*, 3(5):297–302, May 2009.

[224] Brian A. Collins, John R. Tumbleston, and Harald Ade. Miscibility, Crystallinity, and Phase Development in P3ht/PCBM Solar Cells: Toward an Enlightened Understanding of Device Morphology and Stability. *The Journal of Physical Chemistry Letters*, 2(24):3135–3145, December 2011.

[225] Christoph J. Brabec, Martin Heeney, Iain McCulloch, and Jenny Nelson. Influence of blend microstructure on bulk heterojunction organic photovoltaic performance. *Chemical Society Reviews*, 40(3):1185–1199, February 2011.

[226] Ji Sun Moon, Jang Jo, and Alan J. Heeger. Nanomorphology of PCDTBT:PC70bm Bulk Heterojunction Solar Cells. *Advanced Energy Materials*, 2(3):304–308, March 2012.

[227] Tobias Mönch, Peter Guttmann, Jan Murawski, Chris Elschner, Moritz Riede, Lars Müller-Meskamp, and Karl Leo. Investigating local (photo-

)current and structure of ZnPc:C60 bulk-heterojunctions. *Organic Electronics*, 14(11):2777–2788, November 2013.

[228] Antonio Guerrero, Martin Pfannmöller, Alexander Kovalenko, Teresa S. Ripolles, Hamed Heidari, Sara Bals, Louis-Dominique Kaufmann, Juan Bisquert, and Germa Garcia-Belmonte. Nanoscale mapping by electron energy-loss spectroscopy reveals evolution of organic solar cell contact selectivity. *Organic Electronics*, 16:227–233, January 2015.

[229] Wolfram Schindler, Markus Wollgarten, and Konstantinos Fostiropoulos. Revealing nanoscale phase separation in small-molecule photovoltaic blends by plasmonic contrast in the TEM. *Organic Electronics*, 13(6):1100–1104, June 2012.

[230] James B. Gilchrist, Toby. H. Basey-Fisher, Sharon C'E. Chang, Frank Scheltens, David W. McComb, and Sandrine Heutz. Uncovering Buried Structure and Interfaces in Molecular Photovoltaics. *Advanced Functional Materials*, 24(41):6473–6483, November 2014.

[231] M. Pfannmöller, H. Heidari, L. Nanson, O. R. Lozman, M. Chrapa, T. Offermans, G. Nisato, and S. Bals. Quantitative Tomography of Organic Photovoltaic Blends at the Nanoscale. *Nano Letters*, 15(10):6634–6642, October 2015.

[232] Min-Chuan Shih, Bo-Chao Huang, Chih-Cheng Lin, Shao-Sian Li, Hsin-An Chen, Ya-Ping Chiu, and Chun-Wei Chen. Atomic-Scale Interfacial Band Mapping across Vertically Phased-Separated Polymer/Fullerene Hybrid Solar Cells. *Nano Letters*, 13(6):2387–2392, June 2013.

[233] Evan J. Spadafora, Renaud Demadrille, Bernard Ratier, and Benjamin Grevin. Imaging the Carrier Photogeneration in Nanoscale Phase Segregated Organic Heterojunctions by Kelvin Probe Force Microscopy. *Nano Letters*, 10(9):3337–3342, September 2010.

[234] Guozheng Shao, Micah S. Glaz, Fei Ma, Huanxin Ju, and David S. Ginger. Intensity-Modulated Scanning Kelvin Probe Microscopy for Probing Recombination in Organic Photovoltaics. *ACS Nano*, 8(10):10799–10807, October 2014.

[235] Adam G. Gagorik, Jacob W. Mohin, Tomasz Kowalewski, and Geoffrey R. Hutchison. Effects of Delocalized Charge Carriers in Organic Solar Cells: Predicting Nanoscale Device Performance from Morphology. *Advanced Functional Materials*, 25(13):1996–2003, April 2015.

[236] Gang Li, Rui Zhu, and Yang Yang. Polymer solar cells. *Nature Photonics*, 6(3):153–161, March 2012.

[237] Yuhang Liu, Jingbo Zhao, Zhengke Li, Cheng Mu, Wei Ma, Huawei Hu, Kui Jiang, Haoran Lin, Harald Ade, and He Yan. Aggregation and morphology control enables multiple cases of high-efficiency polymer solar cells. *Nature Communications*, 5:5293, November 2014.

[238] Gordon J. Hedley, Alexander J. Ward, Alexander Alekseev, Calvyn T. Howells, Emiliano R. Martins, Luis A. Serrano, Graeme Cooke, Arvydas Ruseckas, and Ifor D. W. Samuel. Determining the optimum morphology in high-performance polymer-fullerene organic photovoltaic cells. *Nature Communications*, 4:2867, December 2013.

[239] Amaresh Mishra and Peter Bäuerle. Small Molecule Organic Semiconductors on the Move: Promises for Future Solar Energy Technology. *Angewandte Chemie International Edition*, 51(9):2020–2067, February 2012.

[240] Richa Pandey and Russell J. Holmes. Graded Donor-Acceptor Heterojunctions for Efficient Organic Photovoltaic Cells. *Advanced Materials*, 22(46):5301–5305, December 2010.

[241] Wolfgang Tress, Karl Leo, and Moritz Riede. Effect of concentration gradients in ZnPc:C60 bulk heterojunction organic solar cells. *Solar Energy Materials and Solar Cells*, 95(11):2981–2986, November 2011.

[242] Janine Fischer, Johannes Widmer, Hans Kleemann, Wolfgang Tress, Christian Koerner, Moritz Riede, Koen Vandewal, and Karl Leo. A charge carrier transport model for donor-acceptor blend layers. *Journal of Applied Physics*, 117(4):045501, January 2015.

[243] Antti Ojala, Andreas Petersen, Andreas Fuchs, Robert Lovrincic, Carl Pölking, Jens Trollmann, Jaehyung Hwang, Christian Lennartz, Helmut

Reichelt, Hans Wolfgang Häffken, Annemarie Pucci, Peter Erk, Thomas Kirchartz, and Frank Würthner. Merocyanine/C60 Planar Heterojunction Solar Cells: Effect of Dye Orientation on Exciton Dissociation and Solar Cell Performance. *Advanced Functional Materials*, 22(1):86–96, January 2012.

[244] Antti Ojala, Hannah Bürckstümmer, Matthias Stolte, Rüdiger Sens, Helmut Reichelt, Peter Erk, Jaehyung Hwang, Dirk Hertel, Klaus Meerholz, and Frank Würthner. Parallel Bulk-Heterojunction Solar Cell by Electrostatically Driven Phase Separation. *Advanced Materials*, 23(45):5398–5403, December 2011.

[245] S. W. Liu, Y. Divayana, A. P. Abiyasa, S. T. Tan, H. V. Demir, and X. W. Sun. On the triplet distribution and its effect on an improved phosphorescent organic light-emitting diode. *Applied Physics Letters*, 101(9):093301, August 2012.

[246] Sebastian Beck, David Gerbert, Tobias Glaser, and Annemarie Pucci. Charge Transfer at Organic/Inorganic Interfaces and the Formation of Space Charge Regions Studied with Infrared Light. *The Journal of Physical Chemistry C*, 119(22):12545–12550, June 2015.

[247] Diana Nanova, Michael Scherer, Felix Schell, Johannes Zimmermann, Tobias Glaser, Anne Katrin Kast, Christian Krekeler, Annemarie Pucci, Wolfgang Kowalsky, Rasmus R. Schröder, and Robert Lovrincic. Why Inverted Small Molecule Solar Cells Outperform Their Noninverted Counterparts. *Advanced Functional Materials*, 25(41):6511–6518, November 2015.

[248] Nobuyuki Otsu. A threshold selection method from gray-level histograms. *Automatica*, 11(285-296):23–27, 1975.

[249] Gwyddion Homepage. http://gwyddion.net/, April 2016.

[250] Alexander Sharenko, Christopher M. Proctor, Thomas S. van der Poll, Zachary B. Henson, Thuc-Quyen Nguyen, and Guillermo C. Bazan. A High-Performing Solution-Processed Small Molecule:Perylene Diimide Bulk Heterojunction Solar Cell. *Advanced Materials*, 25(32):4403–4406, August 2013.

[251] Pabitra K. Nayak, K. L. Narasimhan, and David Cahen. Separating Charges at Organic Interfaces: Effects of Disorder, Hot States, and Electric Field. *The Journal of Physical Chemistry Letters*, 4(10):1707–1717, May 2013.

[252] Andreas F. Bartelt, Christian Strothkämper, Wolfram Schindler, Konstantinos Fostiropoulos, and Rainer Eichberger. Morphology effects on charge generation and recombination dynamics at ZnPc:C60 bulk hetero-junctions using time-resolved terahertz spectroscopy. *Applied Physics Letters*, 99(14):143304, October 2011.

[253] Sylvio Schubert, Yong Hyun Kim, Torben Menke, Axel Fischer, Ronny Timmreck, Lars Müller-Meskamp, and Karl Leo. Highly doped fullerene C60 thin films as transparent stand alone top electrode for organic solar cells. *Solar Energy Materials and Solar Cells*, 118:165–170, November 2013.

[254] Sylvio Schubert, Jan Meiss, Lars Müller-Meskamp, and Karl Leo. Improvement of Transparent Metal Top Electrodes for Organic Solar Cells by Introducing a High Surface Energy Seed Layer. *Advanced Energy Materials*, 3(4):438–443, April 2013.

[255] Rico Meerheim, Christian Körner, and Karl Leo. Highly efficient organic multi-junction solar cells with a thiophene based donor material. *Applied Physics Letters*, 105(6):063306, August 2014.

[256] Christian Willig. *Herstellung und Charakterisierung organischer Dünnschichtsolarzellen basierend auf kleinen Molekülen*. Bachelor thesis, Universität Heidelberg, 2015.

[257] Vera Steinmann, Nils M. Kronenberg, Martin R. Lenze, Steven M. Graf, Dirk Hertel, Klaus Meerholz, Hannah Bürckstümmer, Elena V. Tulyakova, and Frank Würthner. Simple, Highly Efficient Vacuum-Processed Bulk Heterojunction Solar Cells Based on Merocyanine Dyes. *Advanced Energy Materials*, 1(5):888–893, October 2011.

[258] Alhama Arjona-Esteban, Julian Krumrain, Andreas Liess, Matthias Stolte, Lizhen Huang, David Schmidt, Vladimir Stepanenko, Marcel Gsänger, Dirk Hertel, Klaus Meerholz, and Frank Würthner. Influence

of Solid-State Packing of Dipolar Merocyanine Dyes on Transistor and Solar Cell Performances. *Journal of the American Chemical Society*, 137(42):13524–13534, October 2015.

[259] Koen Vandewal, Johannes Widmer, Thomas Heumueller, Christoph J. Brabec, Michael D. McGehee, Karl Leo, Moritz Riede, and Alberto Salleo. Increased Open-Circuit Voltage of Organic Solar Cells by Reduced Donor-Acceptor Interface Area. *Advanced Materials*, 26(23):3839–3843, June 2014.

[260] Kjell Cnops, German Zango, Jan Genoe, Paul Heremans, M. Victoria Martinez-Diaz, Tomas Torres, and David Cheyns. Energy Level Tuning of Non-Fullerene Acceptors in Organic Solar Cells. *Journal of the American Chemical Society*, 137(28):8991–8997, July 2015.

[261] Jeramy D. Zimmerman, Xin Xiao, Christopher Kyle Renshaw, Siyi Wang, Vyacheslav V. Diev, Mark E. Thompson, and Stephen R. Forrest. Independent Control of Bulk and Interfacial Morphologies of Small Molecular Weight Organic Heterojunction Solar Cells. *Nano Letters*, 12(8):4366–4371, August 2012.

[262] Jeramy D. Zimmerman, Brian E. Lassiter, Xin Xiao, Kai Sun, Andrei Dolocan, Raluca Gearba, David A. Vanden Bout, Keith J. Stevenson, Piyumie Wickramasinghe, Mark E. Thompson, and Stephen R. Forrest. Control of Interface Order by Inverse Quasi-Epitaxial Growth of Squaraine/Fullerene Thin Film Photovoltaics. *ACS Nano*, 7(10):9268–9275, October 2013.

7 Appendix

Python script for SRIM input data files

```
file = open("TRIM.dat", "w")
for i in range(1,8):
    file.write("potential comment \n")

file.write("actual comment \n")
file.write("potential comment \n")
file.write("Name  Numb  (eV)  _X_(A) _Y_(A) _Z_(A) Cos(X) Cos(Y) Cos(Z) \n")

xPos = 0
yPos = 0
zPos = 0
#SputterYields:
Yag=17
Yc=2              ─► Enter sputter yields (here: Ga)
Yi=16
Ysi=2
#Sputterionen/Schnitt:
nag=int(200/Yag)
nc=int(200/Yc)
ni=int(200/Yi)
nsi=int(200/Ysi)
#Schichtdicken:
dag=100
dc=600            ─► Enter device stack
di=150                (here: solar cell as discussed above)
dsi=100
#Ordnungszahl FIB-Ion:
o=31
for i in range(1,dag):                          ↑ First layer (Ag with dag)
    for j in range(0,nag):
        file.write( str(i)  +"   "+ str(o) +"   30000   "+ str(xPos*10) +"   0   0   1 0 0\n")
        xPos = i
                                                ↓
for i in range(dag,dag+dc):                     ↑ Second layer (C with dc)
    for j in range(0,nc):
        file.write( str(i)  +"   "+ str(o) +"   30000   "+ str(xPos*10) +"   0   0   1 0 0\n")
        xPos = i
                                                ↓
for i in range(dag+dc,dag+dc+di):
    for j in range(0,ni):
        file.write( str(i)  +"   "+ str(o) +"   30000   "+ str(xPos*10) +"   0   0   1 0 0\n")
        xPos = i

for i in range(dag+dc+di,dag+dc+di+dsi):
    for j in range(0,nsi):
        file.write( str(i)  +"   "+ str(o) +"   30000   "+ str(xPos*10) +"   0   0   1 0 0\n")
        xPos = i

file.close()
```

Figure 7.1: Python script to generate tailored SRIM input data files to simulate FIB milling (comments added in red). With the given script an input file for the simulation of Ga FIB milling of the stack discussed in section 4.3 is obtained. The author thanks Bernd Epding for support with the script.

Calculation of the ion implantation density

This section refers to the results presented in section 4.3.

For the column-wise calculation with Origin the SRIM output files "range.txt" were used, where the number of stopped ions is given for every depth. Here the steps performed in the calculation of the ion implantation (density) are briefly outlined:

- Calculation of the "real" ion number. The numbers given in range.txt are in units of $\#ions/cm^2$, so they had to be normalized to the ion number from the input file.

- Normalization to empirical values. SRIM can only simulate a limited number of ions (99,999), so that we scaled the values from simulation to the values known from FIB exposing solar cell cross sections at our setup. For milling a FIB spot as discussed in the results section with Ga ions, we estimated a need of $\approx 1.3 \cdot 10^6$ Ga ions. For milling a FIB spot with Ne and He ions we multiplied this number with 5 (because $Y_{Ne} \approx 0.2 \cdot Y_{Ga}$) and 100 ($Y_{He} \approx 10^{-2} \cdot Y_{Ga}$) respectively.

- Estimation of the implanted ions. Most of the ions that stopped at a specific depth are not implanted in the sample, but are removed in the ongoing sputter process. We assume here that the ratio of ions remaining in the sample is given by the ratio of FIB spot size and milling resolution. Given the values for the Zeiss Orion, this corresponds to the implantation of every 4th/6th/4th FIB ion into the material matrix for He/Ne/Ga respectively [112].

- Adjusting the milling geometry. We assume that a quarter of the ions of our FIB spot is remaining in the rectangular milled for cross section exposure. With this we obtained the total number of ions implanted at a certain depth.

- Calculation of the implantation density. To calculate the density of implanted ions in the cross section we assumed that they are distributed constant over the lateral 2σ range of the FIB spot. With this the ion implantation density is given by $\#ions \cdot \left(\text{step size} \cdot \pi \cdot (2 \cdot \sigma)^2\right)^{-1}$ with the step size of the range.txt output file.

7 Appendix

CPD maps of bilayer solar cells with MoO_3 coated ITO contact

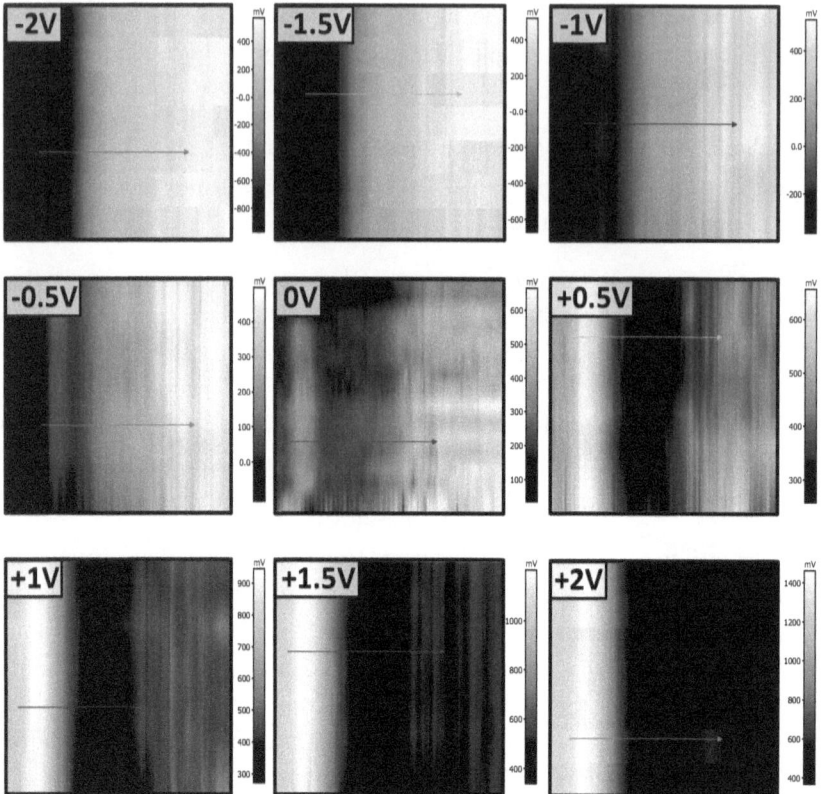

Figure 7.2: All CPD micrographs of the bilayer solar cells with MoO_3 coated ITO contact studied under bias voltage. Spots are marked where the line profiles presented in figure 4.14 were extracted (different colors were used). The image size is $2\,\mu m \times 20\,nm$ for all micrographs shown.

CPD maps of bilayer solar cells with plasma treated ITO contact

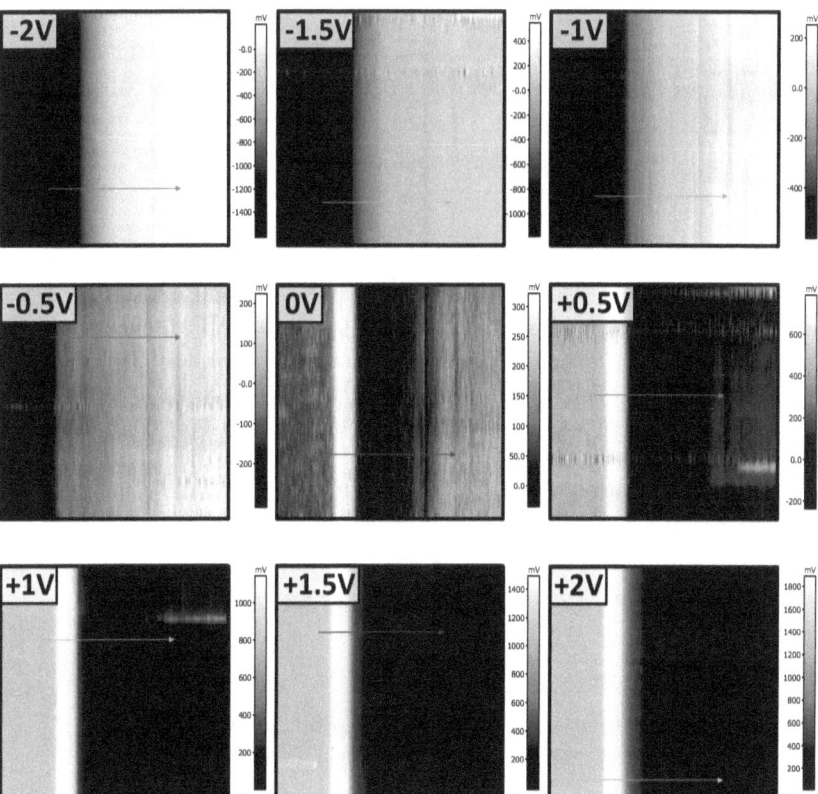

Figure 7.3: All CPD micrographs of the bilayer solar cells with O_2 plasma treated ITO contact studied under bias voltage. Spots are marked where the line profiles presented in figure 4.14 were extracted (different colors were used). The image size is 2.5 µm x 100 nm for all micrographs shown.

7 Appendix

Journal publications, conference presentations, supervised theses

Journal publications

1. Nanova D.*, Scherer M.*, Schell F., Kast A.K., Zimmermann J., Glaser T., Krekeler C., Pucci A., Schröder R.R., Kowalsky W. and Lovrincic R.: *Why Inverted Small Molecule Solar Cells Outperform Their Noninverted Counterparts*. Advanced Functional Materials, 25(41), 6511 - 6518, 2015. *equal contribution.

2. Stolz S., Scherer M., Mankel E., Lovrincic R., Schinke J., Kowalsky W., Jägermann W., Lemmer U., Mechau N. and Hernandez-Sosa G.: *Investigation of Solution-Processed Ultrathin Electron Injection Layers for Organic Light-Emitting Diodes*, ACS Applied Material and Interfaces, 6(9), 6616 - 6622, 2014.

3. Scherer M., Saive R., Daume D., Kröger M. and Kowalsky W.: *Sample preparation for scanning Kelvin probe microscopy studies on cross sections of organic solar cells*, AIP Advances, 3(9), 092134, 2013.

4. Saive R., Scherer M., Müller C., Daume D., Schinke J., Kröger M. and Kowalsky W.: *Imaging the Electric Potential within Organic Solar Cells*, Advanced Functional Materials, 23(47), 5854 - 5860, 2013.

Conference presentations

1. Scherer M., Saive R., Daume D., Kröger M. and Kowalsky W.: *Sample preparation for scanning Kelvin probe microscopy studies on organic solar cells.* Poster; Electronic Structure and Processes at Molecular-Based Interfaces 2013, Weizmann Institute of Science, Rehovot, Israel.

2. Scherer M., Saive R., Daume D., Kröger M. and Kowalsky W.: *Scanning Kelvin Probe Microscopy on cross sections of P3HT:PCBM organic solar cells.* Talk; Tel Aviv University, Group of Prof. Yossi Rosenwaks; 2013; Tel Aviv, Israel.

3. Scherer M., Saive R., Daume D., Kröger M., Lovrincic R. and Kowalsky W.: *Sample preparation for scanning Kelvin probe microscopy studies on organic solar cells.* Poster; JSAP-MRS Joint Symposia 2013, Kyoto, Japan.

4. Scherer M., Bivour M., Türck J., Jenne T., Schell F., Lovrincic R. and Kowalsky W.: *Imaging the electric potential of thin film solar cells.* Poster; DPG Spring Meeting 2015, Berlin, Germany.

5. Scherer M., Jenne T., Schell F., Kowalsky W. and Lovrincic R.: *Imaging the electric potential of organic solar cells.* Talk; MRS Spring Meeting 2015, San Francisco, USA.

7 Appendix

Supervised master theses

- Sebastian Hietzschold, Universität Heidelberg: *Vermessung der Zustandsdichte in organischen Feldeffekttransistoren durch Raster-Kelvin-Mikroskopie.*

- Tobias Jenne, Universität Heidelberg: *Raster-Kelvin-Mikroskopie an Querschnitten organischer F4ZnPc/C60-Solarzellen.*

- Felix Schell, Universität Heidelberg: *Morphology of small molecule organic solar cells from transmission electron microscopy.*

Danksagung

An dieser Stelle möchte ich mich bei Allen bedanken, die mich im Laufe meines Studiums und der Promotion begleitet und unterstützt haben. Insbesondere danke ich

Prof. Wolfgang Kowalsky für das Ermöglichen der Promotion in seiner Arbeitsgruppe am InnovationLab.

Dr. Robert Lovrincic für die wissenschaftliche Betreuung. Ein besonderer Dank auch für das Korrekturlesen dieser Arbeit.

Dr. Diana Nanova, Dr. Janusz Schinke, Sebastian Hietzschold und Patrick Reiser für ihre jeweils besonders großen Beiträge für das Gelingen dieser Arbeit.

Felix Schell für herausragende Ingenieurs- und Gesamtleistungen im Rahmen seiner Masterarbeit.

meinen Bürokollegen und den Kollegen der Analytikgruppe für sehr gute Teamarbeit.

den Kollegen am iL und der iL-Crew um Hildegard Merkle.

Michaela Sauer und Dr. Michael Kröger für großartige Hilfe und wertvolle Beratung.

Sebastian Raupp für großzügiges Asyl im TFT-Büro.

der Hans-Böckler-Stiftung für die Finanzierung meines Studiums.

Allen Saubachern und Bananen.

meiner Familie.
meinen Eltern Brigitte und Erich, meinen Geschwistern Angela und Simon.
Bat-El.
Miri und Gershon.

Herstellung und Verlag:
BoD - Books on Demand, Norderstedt
ISBN 978-3-7412-5152-8

www.ingramcontent.com/pod-product-compliance
Lightning Source LLC
Chambersburg PA
CBHW031623210526
45464CB00004B/1714